D1253661

The
Nature of Change
or the Law of
Unintended
Consequences

An Introductory Text to
Designing Complex Systems
and Managing Change

The
Nature of Change
or the Law of
Unintended
Consequences

An Introductory Text to Designing Complex Systems and Managing Change

John Mansfield
University of South Australia

Imperial College Press

ICP

Published by

Imperial College Press
57 Shelton Street
Covent Garden
London WC2H 9HE

Distributed by

World Scientific Publishing Co. Pte. Ltd.
5 Toh Tuck Link, Singapore 596224
USA office: 27 Warren Street, Suite 401-402, Hackensack, NJ 07601
UK office: 57 Shelton Street, Covent Garden, London WC2H 9HE

British Library Cataloguing-in-Publication Data
A catalogue record for this book is available from the British Library.

THE NATURE OF CHANGE OR THE LAW OF UNINTENDED CONSEQUENCES
An Introductory Text to Designing Complex Systems and Managing Change

ISBN-13 978-1-84816-540-3
ISBN-10 1-84816-540-4

Typeset by Stallion Press
Email: enquiries@stallionpress.com

Printed by FuIsland Offset Printing (S) Pte Ltd. Singapore

Contents

List of Tables

List of Figures

"All things change, nothing is extinguished…
There is nothing in the whole world which is
permanent. Everything flows onward; all things are
brought into being with a changing nature; the ages
themselves glide by in constant movement."

Ovid (43 B.C.–17 A.D.), Metamorphoses, Book 15

1

What Do We Mean by Change?

We exist in four dimensions: the three dimensions of space and the dimension of time. Time only moves in one direction — forward — and it is because of time that we are aware of change. A change occurs only when something is different between two time states; when something moves in space or one of its attributes is altered. This is so much a part of life that we do not even question this definition of change. But there are some things that only form a pattern by changing along the time dimension — music, for instance. A series of changes or patterns in time, be they music or the movement of share prices, are only apparent after the event. It is important to note that change has a binary form, i.e., it happens, or it does not happen. It may be viewed as having the characteristics of a hole: you can have a small hole or a large hole but not a piece of a hole. Likewise, you can have a small change or a large change but not a piece of a change.

A few physicists have challenged the classical view of time flowing through a single universe. Tegmark writes,

"Most people think of time as a way to describe change. At one moment, matter has a certain arrangement; a moment later, it has another. The concept of multiverses suggests an alternative view. If parallel universes contain all possible arrangements of matter, then time is simply a way to put those universes into a sequence. The universes are static; change is an illusion ...."[1]

By this he means what appears to be a change in this universe is actually movement to another, differently configured universe. We move from universe to universe and consider this movement as time. This is an intriguing thought and I will use a similar concept later, that of the design space. However, this multiverse concept provides no insight into the nature of change in this universe other than the conjecture that the probability of moving to a particular alternate universe must vary, some being more probable than others. It is reflected within this single universe by the probability of some arrangement of matter changing to another out of many different, possible arrangements. Whichever way it happens, we perceive this movement as change in our universe.

We, *Homo sapiens*, are a social species and when the first *Homo sapiens* picked up a stone and used it as a tool, complex socio-technical systems came into being. The existence and use of tools has shaped society and society, in turn, has developed tools to fit its evolving needs. These needs have continued to grow so that large, complex socio-technical systems are now found everywhere. Disappointingly, the record of deliberately designing such systems is marked by continual failure and we shall see that this failure occurs because the human designers fail to understand the nature of change.

An example of the unexpected results of change is found in the clearing of trees to make available more agricultural land.[2] This practice has led to rising water tables and increasing salinity that eventually reduces the amount of useable land. Another is the construction of the campanile or freestanding bell tower in Pisa. When the tower was built it was undoubtedly intended to stand vertical. It took about 200 years to complete, but by the time the third floor was added, the poor foundations and loose subsoil had allowed it to sink on one side. Subsequent builders tried to correct this lean and the foundations have been stabilised by 20th-century engineering, but at the present time the top of the tower is still about 15 feet (4.5 metres) from the perpendicular.[3] Along with the unexpected failure of the foundations is the unexpected consequence of the Leaning Tower of Pisa becoming a popular tourist attraction, bringing enormous revenue to the town.

A far more tragic example is the attempt to combat hunger in the Okavango Delta in southern Africa, where periodically the tsetse fly devastated the cattle herds of the native people, often resulting in subsistence-level survival. Western consultants acted to suppress the tsetse fly and replace the local cattle with European-style beef cattle. The increased cattle populations soon overgrazed the pasture, leaving the land an uninhabitable desert.[4]

The creation of unwanted and unexpected effects is a theme addressed throughout this book. It is my view that a major cause of these failures is a general lack of understanding that the people, organisations, hardware, software and any other technology are all part of the same complex system. Not accounting for the interaction among these very different parts of the system after a change is the underlying reason for unpredictable system behaviour. The economist Herbert Simon said many years ago,

> "It is typical of many kinds of design problems that the inner system consists of components whose fundamental laws of behavior — mechanical, electrical, or chemical — are well known. The difficulty of the design problem often resides in predicting how an assemblage of such components will behave."[5]

To understand what I mean, let us consider a traffic jam. A traffic jam is an event that we recognize and it results from the interacting behaviour of many vehicles — cars, bicycles, motorcycles, etc. — but the jam is not evident in the behaviour of any one vehicle; it emerges from the socio-technical interaction of the drivers and their machines. Hence, the effect where the behaviour of the whole system differs from the behaviour of the individual components is known as emergent behaviour. This emergent behaviour is an important characteristic of change and I will talk about it in more depth later. It isn't only technology that fails; Bhopal,[6,7] Chernobyl[8,9] and Three Mile Island[10,11] were all catastrophic failures of socio-technical systems, where human behaviour was an inherent part of the failure. What is more, socio-technical systems are complex systems, and the constant interaction between the system's components creates dynamic effects that are very far from the desired equilibrium. We are all aware of this effect in large information systems. Such systems are subject to this constant interaction: as the users become competent in using the information system, they often see new ways of doing things and dream up new things to do with the information. Additionally, the information system attracts new users with different ideas of functionality. These new concepts change the organisation's (social) processes, structure and its perception of what is required from the information system. So, the technical system's environment changes. To derive the expected benefits, the technical component has to change, thus changing the users' environment. When the social and technical components change in step with each other, we call this process "co-evolution", hence the system's (socio-technical) design co-evolves with the problem space,[12,13] each continually changing in response to the other as the various users' requirements change and the technology changes.

So what controls change? I shall divide the question into two parts, namely what causes an individual component to change and how the propagation of the effects of change through the system affects the system behaviour. The answers to these questions go some way towards answering the key question: How will the system change over time? The concept underlying this book is that the widespread failure to understand or create successful systems (including socio-technical systems) is due to the unrecognised or, more likely, unaccepted fact that all systems are co-evolutionary in nature. I will give a more detailed definition of co-evolution later but I am sure you get the idea. In complex systems change is inevitable and small local changes propagating through the system can cause global changes in system behaviour. Recognition of and accommodation for change is essential to complex, co-evolutionary system design. Designers and implementers of systems demand some form of qualitative prediction of the result of the proposed design in order to create designs that meet the expectations of the system users. This is easier said than done.

Social and technology-based systems are built by lawmakers, engineers, shamans, economists, etc., i.e., almost everyone. System building is goal-directed; there are goals to achieve and requirements to be met. Thus building a system is concerned with plans for the future. These plans are based on predicting the system behaviour at specified times in the future. Indeed, complex systems are built, particularly, to meet an expressed need. But in a large, truly complex system, it is possible for this planned predictability to break down into a morass of unanticipated behaviour so that the need is not met. This unpredictability comes from the propagation of the effects of change through co-evolving components. The effect of a change in one component is often deterministic and generally predictable, but when one component has an influence on one or more of the other components, the effects become complex. In fact the situation is worse than this; the effect of the propagating changes originating from two or more initial changes interfere with one another and alter the outcome. The components' new states apply conflicting pressures for change on each other and can induce them to enter a cyclic or chaotic set of states. These changes to individual components or small groups of components are local in nature but their aggregate effect is evident in the behaviour of the entire system. The holistic system behaviour is unpredictable because it is dependent on a multitude of local changes.

Complex systems are dynamic, i.e., they change over time. Any one component exists in an environment of all the other components in the system. In general, these systems change because each component is influenced by the other components in its environment, as shown in Figure 1.

A simple example is a crossroad where the traffic flow is controlled by traffic lights. If we consider the traffic lights and the vehicles using the junction the system components, then at any given time if the lights change, the behaviour of the vehicles changes and so the system behaviour changes. In this case and generally speaking, there is a perception lag (Lp) between the environment changing and the various components perceiving this change

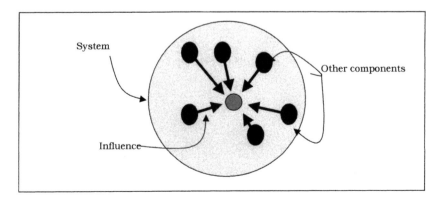

Figure 1. A component in its environment.

in the environment, and a reaction lag (Lr) between the perception of an environmental change and the reaction to that change. In addition, these lags may vary across the components, with one reacting faster than another; $Lp1 \neq Lp2$ and $Lr1 \neq Lr2$. So we will often have a system of many components, each with a different perception lag and reaction lag. If this is so, then shortly after we make a change to one component, some of the other components will have perceived the change and a fraction of these may even have reacted to it. As soon as these components react, they change the environment for the rest of the system. Meanwhile, some of the slower components are now reacting to the earlier change, which also affects the environment. Because of the lags, the components react independently and one small change results in a system that is out of equilibrium. The system and its behaviour are changing over time and interaction is taking place between the components; in short, a typical complex system. I am sure you can relate this to the traffic/traffic lights analogy. Once the process has started, when, if ever, the system will return to equilibrium cannot be predicted by analytical means. However, as we will see later, computer-based simulation shows promise. The rate of change can vary widely but in general, where there are changes incurring small individual costs, you will get rapid rates of change. Changes with large costs tend to happen less frequently. Systems research has shown that for systems as unlike as avalanches in laboratory sand piles,[14] magnitude of earthquakes[15] and the number of deaths due to wars,[16] graphs of the size of the event against the frequency of the event all show a power law. It seems likely that all streams of events in complex systems conform to a power law, even death.

I define the process of change as: influence + decision + transition, but sometimes the decision is not to change and in this case there is no transition.

The Study of Change

Recognition of the importance of change as a phenomenon is not new. Probably the first published research into change was Gabriel Tarde's 1903 book, *The Laws of Imitation*. Tarde, a French judge, wrote about change in social systems from the perspective of a feedback loop between the criminal justice system and criminal behaviour.[17,18] In this approach, he was foreshadowing the better-known systems thinking of Ross Ashby, Stafford Beer and Jay Forrester some 50 years later. Ashby, in his book *Introduction to Cybernetics*, proposed the use of cybernetics, an offshoot of control theory, for the study of changing biological and social systems. He thought that cybernetics was a "theory of machines" but actually, it treats not things but ways of behaving. In keeping with this idea, which works primarily with the general case, cybernetics typically treats any given particular machine by asking not "what individual act will it produce here and now?" but "what are all the possible behaviours that it can produce?"[19]

In effect, Ashby is asking, "What happens when a change is made? How does it feed back into the system?" In the same year, Beer tackled the problem of understanding the organisation of social and socio-technical systems. He based his investigation on insights from neurophysiology, cybernetics and control theory. Initially Beer did not claim to have invented a new model; his aim was to discover and document the persistent pattern of organisations. Nevertheless, in 1981 he developed what he called the Viable System Model (VSM). His long-term goal was to develop a discipline of "management cybernetics" with the VSM as its methodology, and he continued to publish on this theme into the 1990s.[20–27] Forrester[28] introduced the concept of system dynamics, again using ideas drawn from control theory (mainly feedback) to organise information into computer simulation models. His first article based on this work appeared in the *Harvard Business Review* in 1958.[29] Forrester claimed that a digital computer acting as a simulator played out the roles of people in the real system and revealed the behavioural implications of the system that had been described in the model. Given the complexity of human behaviour the early simulations were highly abstract. However, Ashby, Beer and Forrester all accepted that change in one component in a system causes change in other components and, in time, the effects feed back to the initial component. Later, the ubiquity of change in computer-based systems was recognised by Frederick Brooks and in 1975, he wrote about it in his classic book, *The Mythical Man-Month*.[30,31] In the book he says,

> "The only constancy is change itself. Once one recognizes that a pilot system must be built and discarded, and that a redesign with changed

ideas is inevitable, it becomes useful to face the whole phenomenon of change. The first step is to accept the fact of change as a way of life, rather than an untoward and annoying exception."

Brooks perceptively pointed out that the programmer delivers satisfaction of a user's need rather than any tangible product, and both the actual need and the user's perception of that need will change as programs are built, tested and used. Brooks and others emphasise that when the need is poorly articulated and is continually refined as partial solutions are revealed, then change is inevitable. My colleague David Cropley aptly captures this as "the solution defines the need."[32] However, even now, more than a quarter of a century after Brooks published *The Mythical Man-Month*, many socio-technical systems are still not designed with the concept that they might be easily changed, even though the complexity of interaction within and between socio-technical systems may well have increased since 1975. Brooks' point, that the constancy of change in computer-based systems should be accepted as a fact, is clearly still not recognised or found acceptable by a large proportion of system sponsors and builders.

2

Some Definitions

So far we have discussed the basic nature of change but, before discussing it in detail, it is necessary to define a number of the concepts used extensively in this book and to present some concepts to aid the understanding of systems.

Natural Systems

Natural systems are those of the world around us, systems that have evolved without the help of Man. The weather is a very complex natural system that has so many interdependent variables that Man cannot predict its emergent behaviour more than a few days into the future. An ant's nest is a natural system, as is a tree or the ecology of the entire Great Barrier Reef or the human immune system.[33] I am sure you are familiar with many others.

Artificial Systems

Artificial systems are those created by mankind. Some are created deliberately, some accidentally and some just come into being as a result of a person's action and then evolve. In this book I am interested in three kinds of artificial system.

Social Systems

Social systems consist mainly of people interacting one with another, such as a business, a school, a religious system with abstract symbols that are a focus of worship, a football crowd or a local government, all with regulations covering the behaviour of the relevant community. In social systems, the rules governing behaviour may be written down or may be just common knowledge.

Technical Systems

Technical systems are those systems created by people and involving some form of tool, artefact or machine. I include computers and communication systems under the general heading of machines, and substances such as synthetic fabric, buttons and pharmaceuticals under artefacts.

Socio-Technical Systems

Socio-technical systems are systems that combine the two styles of system I have defined very generally in the paragraphs above. Although it is easy to deduce the meaning of "socio-technical", it is still a scientific term and has not yet entered the dictionary. In the middle of the 20th century, the Tavistock Institute created the phrase "socio-technical system" in the context of work-studies. In these studies, they considered technology to be more or less fixed and autonomous, and that the mental and social conditions of human work had to bend to the technical structures. The idea behind the phrase was to emphasise the interaction between people and machines, and this led to what they called time and motion studies. These studies, and the procedures the studies gave rise to, stressed the improvement of efficiency of the human aspect of the socio-technical interaction.[34] Thus, socio-technical is a descriptor for systems consisting of both people and technology, where the people interact with each other and with the technology; the interaction is the important aspect to note. Academics then pointed out that both the social and technical components in a socio-technical system are inextricably linked and must be considered as a unit.[35]

A System

I have used the word "system" a number of times. So what are systems? The definition of the term seems to have changed little over the last 35 years. Stephen Cook has summarised the many variant definitions as

> "Systems can be thought of as comprising elements that in some way work together to create emergent properties; behaviours that are not exhibited from any of the component parts. [...] Importantly, if a group of entities do not synergise to produce emergent properties that differ from those exhibited by the parts, they do not form a system; rather they form a set of parts."[36]

The relationships between the components are very important too. It has become common usage to describe any large, complex system as a "system of systems". Whilst this phrase has some merit, as an indicator that the system so described contains components that are to a greater or lesser extent independent, it offers little more. In this book, the name "system" is given to any hierarchic or networked group of interdependent components that, when regarded as a whole, exhibit a certain behaviour that is not present in any one part but arises from the interaction of the parts. The inherent concept of a system is a set of relationships between components of a group. These relationships may be viewed from various perspectives, perhaps as links between components forming a network structure, or as influences of one component on another creating an environmental pressure for change, or as the degree of inter- and intra-component coupling. The system behaviour that is brought about by this interaction we call emergent behaviour.[33,36–39] Note that where there are many components within a system, this behaviour cannot be predicted from the knowledge of what each component of a system does in isolation. Systems formed from many components fall into two major types, complicated and complex, and we need to understand the difference.

Complicated System

A complicated system is a system of many parts, whose behaviour is that of an automaton, i.e., the individual components and the relationships between the components remain largely unchanged. A mechanical clock is a typical complicated system; another is Brownian movement in a gas. If we take a given volume of gas, we can show that it has the emergent properties of pressure and temperature, properties its component molecules do not possess. I consider this only complicated because the physical laws of thermodynamics can always describe these properties. It is a dynamic system consisting of many interacting molecules but removing a portion of the gas does not change the essential nature of the properties. Compare this with a complex system below.

Complex System

My dictionary defines complex in a traditional, reductionist fashion as "Composed of interconnected parts, compound, composite... a compli-cated assembly of particulars". This bottom-up definition implies that complex merely means composed of interconnected parts. This is quite inadequate. In fact, a complex system may not seem at all complex when viewed as its parts; its complexity cannot be understood when using a linear, reductionist approach. It is the interaction of the many parts that gives a complex system its distinctive behaviour. Robert Axelrod and Michael Cohen in their book *Harnessing Complexity*[40] define complex as an outcome of interaction. They say, "A system is complex when there are strong interactions among its elements, so that current events heavily influence the probabilities of many kinds of later events". This concept of a current event influencing later events is a critical theme in this book.

To the criteria of interaction, the scientist Yaneer Bar-Yam adds the attribute of interdependence. He says,

> "Consider, for example, a person as a complex system that cannot be sep-arated and continues to have the same properties. In words, we would say that complex systems are formed out of not only interacting, but also interdependent parts. Since both thermodynamic and complex systems are formed out of interacting parts, it is the concept of interdependency that must distinguish them."[41]

So if removing a key component from a complex system will change its emergent properties, we can deduce that adding a component to a complex system will also change its emergent properties.

When discussing the concept of a current event influencing later events, it is the propagation of the effects of a change that makes complex system behaviour so unpredictable. Changes to just one sub-component can cause unpredictable interactions that either die away or grow to change global system behaviour. Complex systems can change in multiple dimensions and in the system structure; compare the change from caterpillar to butterfly (decidedly complex) with a cuckoo clock running down (merely complicated).

Complex systems are rarely in equilibrium and hence they must change over time and, as the system is complex, emergent behaviour appears from the system unannounced. One might describe emergent behaviour as similar to a musical chord in that it is quite different to the notes played individually. A local change affecting global emergent behaviour was dramatically demonstrated in September 1998. An explosion in a natural gas plant in Victoria affected much of that Australian state for two weeks. More than 100,000 people were laid off work, and 1.4 million customers went without gas. The economic loss was estimated at almost AUS$1 billion. The impact was felt throughout the country, and the government had to provide more than AUS$100 million in relief.[42,43]

A Component's Environment

As noted in Chapter 1, from the "viewpoint" of one component within a complex system, all the other components are its environment and, of course, the component reacts to the inputs it receives from its environment, so a change in one component may change the environment for many others. The individual component can react to environmental change by invoking a number of purely local rules with deterministic results; but the overall system behaviour can become non-linear and unpredictable. Perhaps the most famous example of this is the movement of the planets. Isaac Newton's inverse-square law of gravity describes celestial mechanics. Using this law, we can write down equations that describe the motion of the Sun or of the planets. If we only consider two bodies, e.g., the Sun and the Earth, and we know their positions and velocities at some start point, we can solve the equations analytically and predict exactly where they will be at any point in time. Unfortunately, when we add a third body to our equations of motion, such as a moon, we can no longer find an analytical solution. Now, the evidence that a problem is not analytically solvable does not, in itself, imply that we are dealing with a complex system, but the behaviour of large, socio-technical systems rarely lends itself to an analytical solution.

It is this fact that is often overlooked, even by the most eminent intellects. At the end of the 19th century Jules Henri Poincaré, a most distinguished mathematician of the time, attempted to determine what is now called the three-body problem, i.e., the collective motion of a sun, a planet and a moon. After 1,500 pages of analysis, Poincaré realised the complexity of the task he had attempted and he noted dryly that it was "a point that gave me a great deal of trouble."[44] Poincaré had encountered deterministic chaos; a situation where each step is still determined by the one before but the system behaviour is chaotic, this phenomenon is also demonstrated by experiments described later in this book. Socio-technical systems have the same "wicked problem", a phrase coined by Horst Rittel and Melvin Webber[45] to describe a problem that has no definitive formulation and no "stopping rule" or defined end-point, so that you can never know when you are done.

Co-Evolutionary Systems

The concept of co-evolution, and its ramifications, is not easy to grasp. Simply put, co-evolution means that as each component in a system changes, it influences many of the other components; in turn these components change to accommodate the initial changes and thus the components that changed initially again experience a pressure to change. This coupling is critical to the concept of co-evolution. There is a view that evolution is adapting to meet one's own needs while co-evolution is adapting to meet each other's needs. I feel this anthropomorphising of the process, implying a form of altruism, is clearly wrong. Individual components only evolve to improve their own fitness; they consider other agents (components) only when those agents influence their fitness. This concept of "influence" is discussed more fully later. A product of co-evolution is not necessarily a system of a higher fitness for a specific emergent behaviour. Change in a co-evolutionary component is simply a local matter and there is no sense of progress of the global system. This idea is most important; evolution in any form is not synonymous with progress but only with change.

Thresholds

When a system component is influenced, even pressured, by its environment it may change its place or its attributes but it may not. The point at which it flips from not changing to changing is called its threshold. There may be many thresholds within the component, thresholds that correspond to its position and each of its attributes. The positional threshold of a billiard ball on a flat table is low and a small physical pressure will cause it to change its position; however, a greater pressure would be required to push the ball up an inclined plane, and an even greater pressure to push it up a vertical wall. In each of these examples the environment is different so the threshold of change is different. In addition to external pressures there can also be pressures within the component when two attributes are in conflict or are linked. In a simple example, when the luxury attribute or the performance attribute of a car is increased, then the linked price attribute will also increase. In some systems this gives rise to what is called a "trade-off", a balancing of attributes against each other to achieve a required end. In complex components with many attributes this can be extraordinarily difficult to achieve.

Propagation

Propagation is the word used to describe the effects of change through time. Because a system is co-evolutionary, one change gives rise to another which gives rise to a third, and so on. What is actually propagating is the wave of effects. In Figure 2 the "colour" attribute of the components changes as the effect of the initial change propagates.

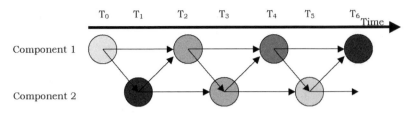

Figure 2.   Change propagating through a two-component system.

Initial Conditions

A major part of the difficulty of predicting the result of a system change is determining the exact value of the attributes of each component at a given time, usually taken as the starting or initial state. This set of values is called the initial conditions; as we shall see later, different sets of initial conditions give rise to different results. Even if the values in two sets of initial conditions are very close, the resultant change may be very different. If one or more of the initial conditions are unknown, or known to an insufficient accuracy, forecasting the result of a change is very difficult. As we are all aware, the weather forecasters with masses of data and number-crunching computers cannot predict the weather with any certainty — their problem is that the atmosphere is a complex system with millions of components and the initial conditions of this system are largely unknown. We will look at the concepts of thresholds, propagation and initial conditions in more detail later.

Design Space

When an object is created it has a number of attributes associated with it. For instance, a chair generally has one, two, three or four legs, it can be blue, red or tweed, it can be large or small, etc. So we can say that a chair has attributes of number of legs, colour and size. For reasons that will become clear later we will rename these attributes and call them dimensions. So the chair exists in a space of, in this limited case, three dimensions.

We can illustrate these three dimensions in a two-dimensional diagram; imagine a box as shown in Figure 3 with the three dimensions

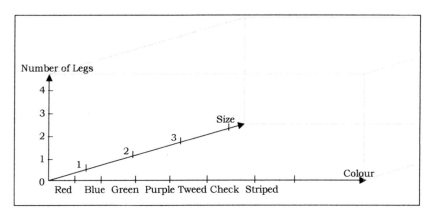

Figure 3. A design space for a chair.

arranged around the three axes of the box. The volume of the box becomes the design space of the chair and encloses all the possible variations of number of legs, colour and size. A component can have many attributes and so the design space is multi-dimensional.

If we take a particular chair, we find it occupies a particular point in the design space as in Figure 4.

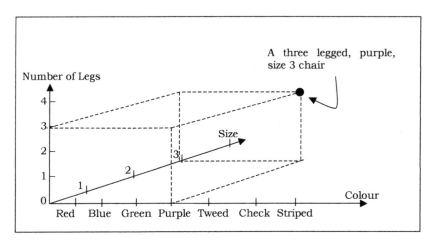

Figure 4. A particular point in the design space for a chair.

3

Failure by Design

Now let us turn our attention to a consequence of the co-evolutionary nature of complex systems, the propensity for human-designed systems to do things that the designer did not expect.

Socio-Technical Design Failure

The record of deliberately designing socio-technical systems is marked by continual failure. Socio-technical "designs" often end up causing the opposite effect to that which was intended; I mentioned tree clearing,[2] the Leaning Tower of Pisa[3] and the Okavango environmental/agricultural tinkering[4] in Chapter 1. There are plenty of other documented examples where computer-based socio-technical systems fail to live up to their promise.[46,47] All the characteristics of complexity, co-evolution and continuous change are present in large, computer-based socio-technical systems so I will take them as my main illustration. This complexity is counterproductive and has emerged as one of the key challenges facing modern businesses, and even the best technical minds in the world are struggling to come to terms with the rapid change in modern computer systems.

Computer-based information systems are particularly fallible; they often fail to live up to their designer's promises, a point confirmed in a report by Jim Johnson that he called "Chaos: The Dollar Drain of IT

Project Failures".[46] Top business people are worried about this fallibility. At the World Congress on IT in Adelaide, Australia, in March 2002 IBM Global Services chief executive Doug Elix was quoted as saying, "[Infrastructure] has grown to the extent where fixing problems is often impossible."[48] At the same conference, Commonwealth Bank managing director David Murray blamed the US computer industry for "single-handedly wrecking the global economy."[49] They were both referring to the complexity and failure of the computer-based system's design.

Moreover, they have good reason for their pessimism. In 1995 Johnson surveyed 365 American companies and found only 12% of 3,682 computer projects were finished on time and budget, 31% of projects were cancelled before completion, and 53% overran their budget and had impaired functionality, so only 442 out of 3,682 were judged successful! Few countries have gone unscathed by a major failure. In the US there was the Denver Airport luggage-handling system[50]; in France, SNCF's SOCRATE reservation system[51]; in the UK, London Ambulance Service's despatch system[52]; and in Australia, where this book was written, the AUS$150 million failure of WestPac's CS90 banking system,[53,54] the AUS$91 million RMIT/PeopleSoft debacle[55] and the AUS$409 million National Australia Bank software write-down.[56] An even larger blunder was the failure of Cisco's inventory system. In May 2001, Cisco Systems (CISCO) announced the largest inventory write-down in history: US$2.2 billion was erased from its balance sheet for components it ordered but could not use. What made it more embarrassing were the waves of prior publicity about Cisco's brilliant integration of its vast information systems.[57] In their book, *Management Information Systems*, Kenneth and Jane Laudon comment, "As many as 75% of all large [information] systems may be considered to be operating failures."[58]

The skill sets of the practitioners may be to blame but it seems more likely that the majority are ignoring the fact that they are building a socio-technical system and that in such systems change is continuous and interactive. It is not uncommon for system developers to comment wearily, "Not another change, I do wish the users would make up their mind." By making this comment they clearly have a mindset that identifies their system development with a mechanistic building process rather than it being only a part of a much larger, co-evolutionary socio-technical system. The common view of requirements and systems analysis is that analysis determines what a system should do as if it were clear-cut; it is my belief that a more practical view is that systems analysis is like archaeology where some artefacts are exposed but much is left to be found at a later stage. Similarly, system design, particularly socio-technical system design, is thought of as engineering, as if it were governed by physical laws, but experience dictates that it is closer to biological evolution in a constantly changing environment.

In view of the demonstrated difficulties of achieving pre-stated ends by constructing socio-technical systems, there is an urgent need for insight into the design of complex socio-technical systems that can evolve without breaking. It is possible that the solution hinges on the context-sensitive nature of change in co-evolutionary systems. Showing that socio-technical systems can be modelled as co-evolutionary systems and simulated on a computer, the dynamic process inherent in the creation of large computer-based systems can be demonstrated.

In a co-evolutionary system, any component's environment consists of all the other components in the system, so it is clear that change in any one component affects that component's environment. This effect will then feed back into the original component; given the multiple interactions, the behaviour of these systems is, for the most part, unpredictable, and even the designers do not know what the system will do when it is "completed". Herbert Simon's observations, made as long ago as 1968, acknowledged this concept:

"The research that was done to design computer time-sharing systems is a good example of the study of computer behavior as an empirical phenomenon. Only fragments of theory were available to guide the design of a time-sharing system or to predict how a system of a specified design would actually behave in an environment of users who placed their several demands upon it. Most actual designs turned out initially to exhibit serious deficiencies, and most predictions of performance were startlingly inaccurate. Under these circumstances the main route open to the development and improvement of time-sharing systems was to build them and see how they behaved. And this is what was done. They were built, modified, and improved in successive stages. Perhaps theory could have anticipated these experiments and made them unnecessary. In fact it didn't, and I don't know anyone intimately acquainted with these exceedingly complex systems who has very specific ideas as to how it might have done so. To understand them, the systems had to be constructed, and their behavior observed."[5]

Here again is the archaeological analogy. Simon suggests that the implementers scraped the surface, found some things that worked and built on those things, discarding the mistakes. Despite the findings published by Simon and his colleagues, computer-based socio-technical systems still seem to be designed mechanically with all components interwoven, and changing one component requires changes to all the others. This technique is inadequate for systems today; now they require far greater flexibility, modularity and less coupling, or influence, between the modules.

Back in 1969, Butler Lampson commented that

"If a system is to evolve to meet changing requirements, and if it is to be flexible enough to permit modularisation without serious loss of efficiency, it must have a basic structure that allows extensions not only from a basic system but also from some complex configuration that has been reached by several prior stages of evolution. In other words, the extension process must not exhaust the facilities required for further extensions. The system must be complexly open-ended, so that additional machinery can be attached at any point."[59]

If this concept was recognised nearly 40 years ago, why do systems still fail? Perhaps because designers believe they have the foresight to design for the future needs of users, whereas, in fact, it is the system architecture that must facilitate change in the infrastructure and in the components. More importantly, the designers, users and system sponsors must all recognise that they too are part of a co-evolving system and that change must be expected. If computer-based, socio-technical system designs are based on the concept of co-evolution, they can grow over time and develop a complexity that no designer could have envisaged before implementation. These systems build continuously on previous experience and need never be thrown away.

For computer-based systems, the rate of environmental change is increasing so if designers ignore the co-evolutionary essence of socio-technical systems, they will continue to be unsuccessful. The 2003 *CHAOS Chronicles* report shows some improvement and some deterioration since the first *CHAOS* report in 1994:

> "Project success rates have increased to just over a third or 34% of all projects. This is a 100% plus improvement over the 16% rate in 1994. Project failures have declined to 15% of all projects, which is more than half the 31% in 1994. Challenged projects account for the remaining 51%."

This looks like progress in system design and implementation.

> "However the report does not show all good news. Time overruns have significantly increased to 82% from a low of 63% in the year 2000. In addition, this year's research shows only 52% of required features and functions make it to the released product. This compares with 67% in the year 2000."[47]

This report indicates that from the perception of those initiating projects, 34% were successful, 15% were definite failures, and the remaining 51% were unsuccessful in one way or another. This is not a record of outstanding competence. From the improvement in the number of projects judged successful, one may infer that project sponsors have become more pragmatic and adherence to the budget has become the pre-eminent parameter in judging success. However, this is at the expense of either time overruns or the removal of required features and functions. In the latter case, it is unlikely that the users will agree that the project was a success. It is probable that when these systems go into service they will continue to co-evolve with the needs of the users. So-called system "maintenance" becomes this process of continued co-evolution rather than the normal concept of maintenance — that of fixing glitches due to wear and tear.[50,60,61]

If the majority of computer-based socio-technical systems fail to meet the expectations of their sponsors, perhaps this is due to their architecture. There is apparently no research comparing the success of large systems that were built from the specification without change with those that underwent change during implementation. This is probably because no examples of systems with unchanging requirements exist.

The phenomenon of change during implementation was apparent from the earliest systems — Brooks based his comments in *The Mythical Man-Month*[30] on change in large systems built in the 1960s and 1970s. We can therefore assume that all modern systems have a co-evolutionary implementation. By taking this co-evolutionary perspective, the design of complex socio-technical systems can be viewed as the series of evolutionary moves that the individual components make over time. Quite frequently, system changes turn out either to have been unanticipated, or to have unexpected and unintended consequences. This is because the socio-technical systems are complex, co-evolutionary systems. In the context of social planning, Rittel and Webber[45] called the co-evolutionary problem "wicked" because it has no definitive formulation and no stopping rule, therefore you can never know when you are done. It should be noted, however, that a wicked problem is not necessarily co-evolutionary. I have shown above that many large information technology systems fail to achieve the goal of their sponsors. Yet, information technology systems are but one subset of the countless large socio-technical systems that have failed to a greater or lesser extent. Other subsets are spacecraft (e.g., Columbia[62]), military systems (e.g., the Collins submarine[63]) and industrial systems (e.g., Bhopal,[6,7] Chernobyl[8,9] and Three Mile Island[10,11]). There is a virtually endless list of socio-technical systems devised and implemented by humans but with a record of not meeting their sponsor's expectations and the system engineer's predictions. These failures occur because socio-technical systems are inherently complex, co-evolutionary systems, and when such systems reach a particular level of complexity, they become as unpredictable as eddies in a stream.

4

Influence, Boundaries and Structure

As I have noted before, a change is defined as influence + decision + transition, but decisions within socio-technical systems can be very difficult. Described below are two decisions that might go to one option or another. For simplicity, the examples have only two options but in reality, almost all decisions have multiple options.

Examples

Case 1

Influences

My daughter asks me to pick her up from the cinema after dark. Should I agree?

Yes, because

- It is my moral obligation as a father to assist my child.
- She would get home safely and swiftly.
- It would alleviate her mother's concern for her safety.

No, because

- I would have to miss a favourite TV programme.
- I would have to leave my warm home on a cold night to drive across town.
- I would have to forego my wine with dinner.

Decisions

In this case, the probability is that I would agree. However, the decision could go the other way if I have urgent work to finish and this is the fifth time this week she has asked me to pick her up.

Transition

In this case the transition is the act of getting into my car at an appropriate time to be able to meet my daughter as she leaves the cinema.

Case 2

Influences

I regularly drive to and from work along a main road; due to lobbying by a road safety group several sets of traffic lights have recently been installed. The traffic lights increase the transit time along the road by five minutes. Should I continue to use the road rather than divert via several smaller suburban roads?

Yes, because

- I am aware that the decision to add traffic lights was a safety issue and it is my social responsibility to accept the delay.
- In my overall journey time five minutes is a small variation.
- I'm too lazy to work out a route through the neighbouring suburbs.

No, because

- I am aware that the decision to add traffic lights was a safety issue and I am irritated by minority groups dictating my behaviour.
- I am always running late and every second helps.

- I am unmoved by the potential consequences of driving quickly through narrow suburban roads where I will encounter children, old people and pets on the road.

Decisions

In this case, the probability is that I would continue to use the main road. However, if I was late for an appointment and the traffic on the main road was very heavy I might ignore the danger to myself and other road users and use the side roads.

Transition

I continue the current behaviour in the face of a changed environment.

New influences after the change

After the change has taken place, the new circumstances (component environment) then influence other actors (components) in the environment. In Case 1 my daughter asks me to give her friends a lift, requiring me to take a long and tedious detour. On the way back, because I am unprepared, I run out of petrol and there are no petrol stations nearby, I have to walk for miles to find a petrol station, leaving my daughter in the car. On my return to the car, I find a gang of youths banging on the sides of the car, terrifying my daughter. When we finally arrive home very late, we find my wife distraught with worry. There is a bushy tree of possible options that propagate from the initial decision; one example is illustrated in Figure 5.

A similar tree containing benign and disastrous end states can be constructed for Case 2. The propagation of the initial decision to one of a number of unpredicted states is the source of the uncertainty of change in any complex system.

Influence

A good definition of influence is this one from the *Macquarie Dictionary*: "To move or impel to, or to do, something".[64] Causal networks, which

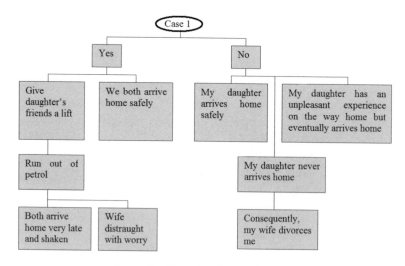

Figure 5.   Tree of options and results.

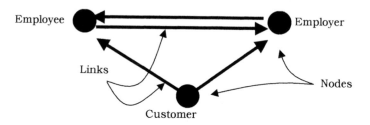

Figure 6.   A directed graph.

show how influence is deployed, are often depicted as a directed graph as illustrated in Figure 6; it is called directed because the arrowheads on the links show the direction of the influence. This, of course, means that if two components each influence the other, the diagram will show two arrows, one pointing in one direction and the other pointing in the opposite direction. We will often meet this type of diagram in the following chapters.

An influence contributes to a decision, the influence may be necessary to reach a particular decision, but it may not be sufficient. In a chess game, the set of positions of your opponent's pieces influences what your next move is but does not control it. As an agent with choice, other influences can be brought to bear; you can make a number of different moves. The actual move is dependent on the "aggregate" of the influences and the boundary of the system.

In a socio-technical system many relationships may be represented concretely, albeit qualitatively; an employee's influence on his supervisor's behaviour is considerably less than the supervisor's influence on the employee's behaviour. In a model system the magnitude of the influence in these relationships can be defined by rules or mathematical functions, or given an empirical value.

For a change to occur within a socio-technical system, all the influences on a component need to be "aggregated". The influences for the change must then outweigh the influences against it. As we noted before, in such a system all the components lie in an environment that consists of all the other components. The effect that changes in those other components have on an individual component is dependent on the influence they have on that individual component. Thus in this context, influence can be defined as the magnitude of the pressure for change one component exerts on another.

Boundaries

To define a system within which this influence operates we need a boundary. The first consideration when studying a system is where the boundary should be drawn. Where the boundary is drawn when creating a model or description of a complex system is quite arbitrary. Keep in mind that it is drawn to abstract a manageable view of the system from the messiness of the real world. In the real world, of course, boundaries are permeable but in a model or description, once a boundary is drawn, the influences from outside the boundary are assumed negligible and hence neglectable. Therefore any influence that is considered to have an effect on the system must be included within the system boundary by definition. The problem lies in deciding how much effect is "an effect". Influences from outside the boundary may fall off with distance, like light or gravity over astronomical distances.

In Figure 7, where do we draw the boundary, at the inner or the outer circle?

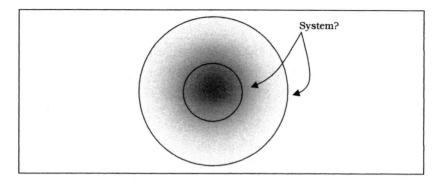

Figure 7.   Graduated influence.

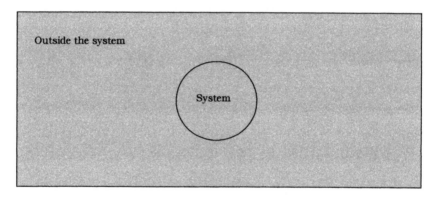

Figure 8.   Pervasive influence.

Influences can be unvarying both inside and outside the system boundary, as is the effect of gravity on a system on the surface of the Earth (Figure 8). Such influences can have a major effect on the system but, because they are all pervasive and apparently immutable, they are ignored to simplify the abstract view. This may have a catastrophic effect when the external influences change. When farmers clear trees to provide a greater fertile acreage and ignore the fact that they are lowering the water table, salinity rises through their soil and makes all of it infertile. It becomes even more difficult when there are multiple influences of different strengths.

Take, for instance, the light-sensitive system in Figure 9 with influencing light sources at varying distances. Since the power of the light falls off

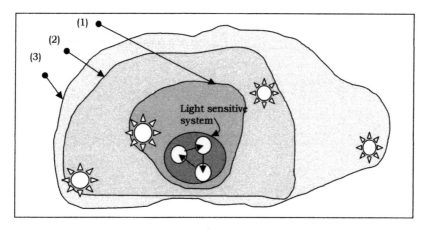

Figure 9. Multiple influences of different strengths.

as an inverse square of the distance, is (1) the system boundary or (2) or perhaps (3)? The answer is that the position of the boundary depends largely on the purpose of the model and what you wish to study or design. But do not forget that influences outside the boundary will be ignored.

Organisation or Network Structure

Mathematicians portray a network of influences as a directed graph and it is this network of influences between the components that determines the organisation of a complex system. These networks can vary in structure from regular to random, and where they fall in this "spectrum" determines their properties. The "small world" structure has particularly interesting characteristics. Change in the organisation of a complex system comes from the creation, destruction, strengthening and weakening of these influence links.

Regular Networks

As shown in Figure 10, a regular network consists of a symmetric arrangement of a number of components. In the diagram below the regular network consists of a circle of twelve nodes and twenty-four links connecting adjacent and alternate nodes.

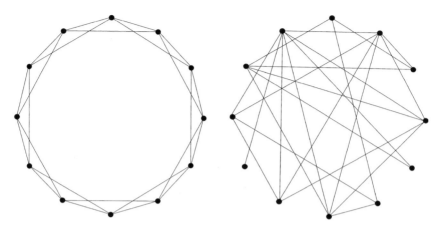

Figure 10.   A regular (*left*) and a random (*right*) network.

Random Networks

A random network consists of a set of nodes connected at random. More formally, a random network is formed when a network is connected such that a link between two nodes is formed with a probability p, duplicate links and self-connections being excluded. By varying the probability, the transition from regular to random networks can be studied.[65] Paul Erdös and Albert Rényi coined the concept of random networks in a paper published in 1959[66] and since that time mathematicians have published many papers on the topic of random networks.

Unfortunately the random graph as defined by Erdös does not relate well to real-world networks such as socio-technical networks, the World Wide Web, electrical power systems, etc.[67] This discrepancy led Duncan Watts to study the characteristics of an arrangement that he called a small-world network.[68,69]

Small-World Networks

A small-world network is part-way between a regular network and a random network, and includes many local links between groups of nodes with a few links to other similar groups. In other words, there are many clusters of nodes linked by an occasional long link as shown in Figure 11.

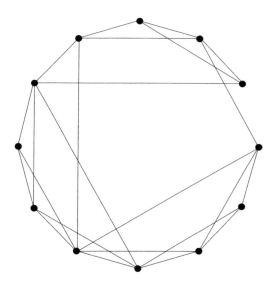

Figure 11. A small-world network.

Duncan Watts and Steven Strogatz[68] used this type of ring lattice to demonstrate the characteristics of networks with differing forms of connection between nodes, with particular reference to small-world systems.

In a complex, co-evolutionary system the influence connections (links) form the network. The path length between any two nodes across these networks can be defined as the number of nodes traversed to go from one specific node to another specific node. Thus, the path length determines how quickly a change in one node is propagated to another across the network. In the case of regular networks, the average path lengths are long, because there is no direct route from one side of the network to the other, so propagation is slow. With random networks, the average path length scales logarithmically with the number of nodes and can be calculated as the average number of neighbours a node has, hence networks with many nodes have increasingly long average paths and the average speed of propagation is again slow. This characteristic of long average path lengths departs from that shown by naturally occurring complex networks.

Small-world networks, in contrast, have low average path lengths as the few long distance links permit "short cuts" across the network, and so propagation of change across the network is fast. Later I will discuss a validation of the hypothesis that systems with a small-world structure can rapidly change their emergent behaviour.

Layered and Hierarchic Aspects

When we examine evolutionary systems such as ecologies and business organisations, we notice that in addition to having boundaries they are divided into layers. For instance, the layers in a building include the structure (foundations, outer walls and roof), the services (water, electricity, telephone, etc.), temporary walls, furniture and people. In an ecology, there are forests, deserts, trees, shrubs, mammals, birds, annual flowering plants, beetles and butterflies. In each layer, we can also notice that the local ecosphere is on a different scale from those above or below it. The forest and the desert respond to long-timescale climatic changes but their response to the death of a flower or butterfly is imperceptible. The death of a tree has a local effect but the effect on the forest is small. In addition to scale, these layers also differ in their rate of change; in general the larger the scale, the slower the rate of change. Thus, in this type of system model, scale, layering and rate of change become interchangeable concepts.[38,70]

The interaction between layers is usually small until a critical threshold is reached, after which a small change in one layer can cause a change in upper or lower layers; for example, we can push more and more desks into an office but there comes a time when to get more into the room, space-dividing walls must be moved. This may be the precursor to a cascade of changes and can be likened to a phase change where dropping the temperature of water by one degree may cause it to change to ice. Scientists have found corollaries for layers in information systems, economics[71] and business.[72,73]

It seems to me that socio-technical systems that have a formal design are designed in the way that architects used to design buildings a hundred or more years ago[74] — the space plan, structure, etc., were all tightly interwoven and changing one required changes to all the others. A great opportunity will open for the person who can determine just what the layers are in, say, a computer-based system's architecture, so that we can achieve the same flexibilities as in ecologies and buildings.

Systems have a Past and a Future

Change, by nature, is a temporal phenomenon. There has to be a time before the change and a time after the change for it to be perceived. From very different fields and perspectives, both Herbert Simon[5] and Stephen Jay Gould[75] agree that change is contingent or history-dependent. The current condition of a system is entirely contingent on the long series of past conditions and the changing environments that shaped them, and future change is contingent on the current condition of a system.

Newton's First Law states that "Every body continues in its state of rest or of uniform motion in a straight line unless acted on by a force". This implies that any closed system in internal equilibrium will not change, but the keyword is closed; no system that we know of is unaffected by other systems. This point was discussed in the section on system boundaries. Whether the system in question is a galaxy, a computer system or a bacterium, it exists in an environment of other systems, i.e., it is an open system. These other systems produce streams of events that impact the system of interest and if their impact is sufficiently severe, they will change the system. Remember that change is influence + decision + transition. For example, complex computer-based systems exist in a dynamic environment; almost everything that defines or affects them is changing at an increasing pace.

Nevertheless, there is a significant cost in throwing away a system, creating new systems and even in starting up and shutting down a system. Add to this the need for many computer-based systems to be continuously available; in defence, banking, medical, even social security payment systems, losing the system's availability has a heavy financial, social or political effect. Hence, systems need to change on the run, i.e., they must evolve in response to external factors.

Layers of Change

The layer concept of complex systems also applies to evolutionary change; such change happens at different rates for different components. For a planet's structure, the timescale is many millions of years, and the rate of drift of the continents is about the same as the growth of your fingernails.

Hence, mountains are created at a very slow rate. The effect of weather erosion on a mountain is also slow but much faster than continental drift. Similarly, in socio-technical systems some components change faster than others due to the varying rate of impact of the events affecting them. In opposition to the pressure of the stream of events is the cost of change; there is a disturbance cost to any change. Changing a mainframe computer has a huge disturbance cost and so requires an equally large pressure for change. Discontent with a user interface may be minor, in terms of pressure for change; although, if enough minor pressures accumulate, together they can create a strong pressure for change.

Change in a system can happen at many different levels, different rates and with different costs. A small change is often one with a small cost. Small changes typically happen at a fast rate because they only cause a small disturbance. Conversely, small changes sometimes have large, unforseen ramifications in complex systems and many small changes over a period of time can have an impact disproportionate to the change by virtue of the feedback effect. Classically, modifying a line of program code may be a trivial change and correct in its local context, but it may cause the entire system to fail.

To show how an accumulation of small changes may cause a system to change, we will carry out a couple of thought experiments, i.e., we will only conduct the experiment in our mind.

We know that change occurs when the pressure for change exceeds the resistance to change. An external change may cause irresistible pressure for change and hence lead to instant change, or a series of tiny external changes may accumulate slowly until the pressure they generate rises above the system's resistance to change.

The apparatus for the experiments is shown in Figure 12: a thin card is placed on four strong supports; the card is stiff enough to remain flat even though it is supported only at the corners.

Experiment 1. Pour sand evenly onto the card, as shown in Figure 13. As the sand builds up it flows over the edge but accumulates in the centre until it gives way and spills the sand on the floor. This experiment takes place at room temperature, say 20°C. Would a different ambient temperature give a different result? Probably not!

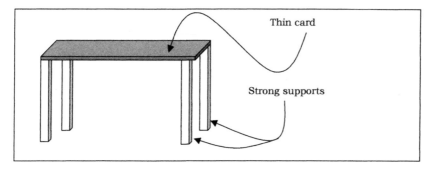

Figure 12. Thought experiment apparatus.

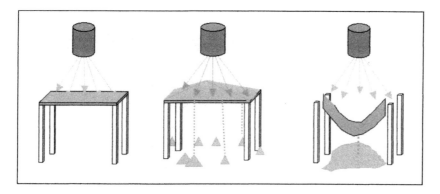

Figure 13. Thought experiment.

Experiment 2. Now we replace the sand with snow. The card should collapse with a similar weight of snow but now the temperature will be relevant. If the temperature is at +2°C the snow will melt (change) and wet the card, weakening it (change) and causing it to collapse at a lower applied weight. But if the snow melts it will flow off the card and will not build up, so even the weakened card will retain sufficient stiffness to resist change (no change). However, if the temperature is −2°C the snow will act in a similar manner to the sand and will build up on the card. But, unlike sand, the snow in contact with the card will melt (change) under the pressure of the overlying snow and seep into the card where it will freeze (change). The frozen card is stiffer than the dry card so it will continue to resist the weight of snow equivalent to the weight of the sand that caused the dry card to collapse (no change). We can see that even a very simple system, in this case a supported card, will react differently to the same increasing external pressure for change, i.e., the weight applied to the centre of the card. In this case it is clear that the system reacts differently because the system environment is different, although it is superficially similar and the pressure for change due to an applied weight is the same. These experiments were carried out with a deliberately simple system but even here, we have ignored a number of possible environmental variables. You can probably think of a few yourself.

5

Change in Complex Systems

As we saw at the end of the previous chapter, even simple systems can be hard to understand; now there are man-made systems, such as large, complex computer-based systems that have moved beyond the detailed understanding of a single man or woman. We need to stand back and try to understand the principles of change in the complex system embodied by large, modern, computer-based systems. In this context, the comments of the biologist Lewis Thomas in his essay "On Meddling" are enlightening:

> "When you are confronted by any complex social system, such as an urban centre or a hamster, with things about it that you're dissatisfied with and anxious to fix, you cannot just step in and set about fixing with much hope of helping ... You cannot meddle with one part of a complex system from the outside without the almost certain risk of setting off disastrous events that you hadn't counted on in other, remote parts. If you want to fix something you are first obliged to understand ... the whole system ... Intervening is a way of causing trouble."[76]

This insight is not novel. The authors of the 1662 edition of the *Book of Common Prayer* recognised that the insight was ancient more than 300 years before Thomas did. They start the Preface by saying,

41

"It hath been the wisdom of the Church of England, ever since the first compiling of her publick Liturgy, to keep the mean between the two extremes, of too much stiffness in refusing, and of too much easiness in admitting any variation from it. For, as on the one side common experience sheweth, that where a change hath been made of things advisedly established (no evident necessity so requiring) sundry inconveniences have thereupon ensued; and those many times more and greater than the evils, that were intended to be remedied by such change."[77]

Similarly, in the context of organisational change Sterman commented,

"In 1516, Sir Thomas More wrote in Utopia about the problems of policymaking, saying, 'And it will fall out as in a complication of diseases, that by applying a remedy to one sore, you will provoke another; and that which removes the one ill symptom produces others.'"[78]

These last two comments are reminiscent of system debugging. By illuminating some small part of the concept of change in complex, computer-based systems, this book may turn the ramifications of change to our advantage rather than provoking other "sores". They also remind me of the programmer's song:

> "99 little bugs in the code,
> 99 little bugs in the code,
> fix one bug, compile it again,
> 101 little bugs in the code…" *Anon.*

It is the nature of co-evolutionary systems that a change in one component has an influence on the other components, the effects of which may feed back to the original component. In biology, Richard Dawkins calls this the biological "arms race".[79] For example, as the cheetah becomes faster the antelope becomes more agile, so the cheetah has to become faster still and so on. The arms race between the cheetah and the antelope is, of course, not that simple; many other biological and environmental components affect the outcome and they are then affected in turn. This gives rise to a high degree of uncertainty in the effect of the cheetah getting faster.

This uncertainty is the major difficulty in the prediction of the effect of a change in a co-evolutionary system, and if the result is uncertain then design and control of such systems are made more difficult. Thus, "uncertainty management is a fundamental unifying concept in analysis and design of complex systems".[80]

For the purpose of illuminating the process of change and effect and reducing the uncertainty, I have devised a protocol or set of rules for change.

Fitness and Change

The fitness of a system is a measure of whatever combination of attributes gives the most "benefit" in the current circumstance. This could be survival to reproductive age (biology), financial profit (business) or popularity (music groups). Fitness is a balance between two or more attributes, a trade-off of cost against benefit. For fitness to exist as a concept every system and component in that system has to have an inherent fitness algorithm that works out its fitness, this algorithm may be tangible in a model but intangible in the real world. The attributes and mechanism of a fitness algorithm are a critical part of the protocol for change.

Referring to the concept of fitness in a multi-dimensional universe, Herbert Simon comments,

"To deal with these phenomena, psychology employs the concept of aspiration level. Aspirations have many dimensions: one can have aspirations for pleasant work, love, good food, travel, and many other things. For each dimension, expectations of the attainable define an aspiration level that is compared with the current level of achievement. If achievements exceed aspirations, satisfaction is recorded as positive; if aspirations exceed achievements, there is dissatisfaction. There is no simple mechanism for comparison between dimensions. In general a large gain along one dimension is required to compensate for a small loss along another — hence the system's net satisfactions are history-dependent, and it is difficult for people to balance compensatory offsets."[5]

In dealing with the concept of fitness Simon confirms there is no simple mechanism that can define the fitness algorithm so that it includes the effects of the necessary "trade-offs", thus the fitness algorithm is of necessity complex. This passage also reinforces the contention that the outcome of the fitness algorithm is history-dependent.

A significant insight can be drawn from the idea that change in a component can be considered as movement on a fitness landscape, an idea due to Sewall Wright[81] and extended by Stuart Kauffman.[82] To illustrate the idea of a fitness landscape, consider another thought experiment. Imagine a component with two attributes or dimensions, A and B, mapped onto a design space in which a two-dimensional plane represents all possible combinations of the two dimensions. Any point (a, b) on the plane represents a specific combination of the two dimensions, for example the black dot in Figure 14.

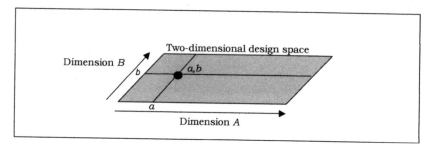

Figure 14.   A two-dimensional design space.

This component exists in an environment and, within that environment, a value can be calculated for the "fitness" of any combination of A and B (Figure 15).

This value can represent performance, cost, speed, political or social benefit, etc., and is dependent on the fitness algorithm of the component in its current environment. Stuart Kauffman equated it with the ability of a species to survive from Herbert Spencer's phrase the "survival of the fittest".

Now imagine that at each combination of A and B there is a line at right angles to the design space plane, of a height proportional to the fitness of that combination in the current environment (Figure 16).

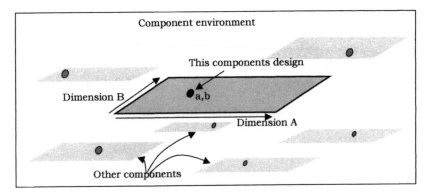

Figure 15. A two-dimensional design space in its environment.

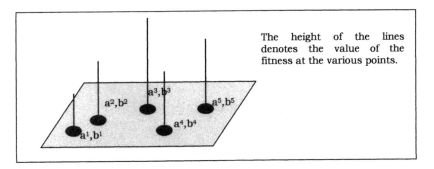

Figure 16. Fitness magnitude at different points on the design space.

The tops of these fitness lines form an undulating, three-dimensional surface that Wright[81] called the fitness "landscape" because of its similarity to a topographical map (Figure 17).

Now imagine a second component within the same environment with its own fitness landscape. If these are the only two components within the system, each forms the external environment for the other. Thus when one changes due to internal or external pressures, it changes the environment of the other and hence the other's fitness landscape. For example, when the cheetah learns to run faster, the antelope's environment and hence its fitness landscape changes, and the antelope's fitness is reduced even though the antelope has not changed. This is illustrated in Figure 18.

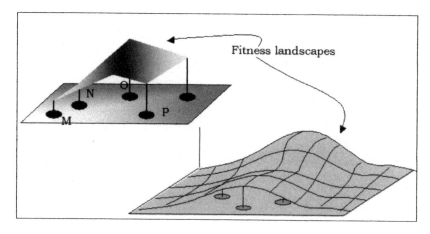

Figure 17.   Fitness landscape.

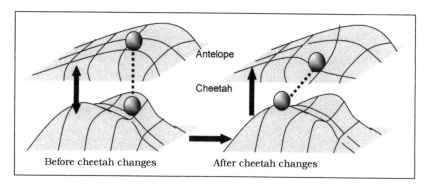

Figure 18.   Deformation of a fitness landscape as linked fitness landscape changes.

Now consider a system with many components. There is a multi-dimensional, interlinked fitness landscape within such a complex system and that landscape is altered by any change of any component. Hence component A, by doing nothing, can become less or more fit as a consequence of other components changing. In turn, this change in fitness of component A may produce a pressure for change on component B.

As a practical example, Figure 19 shows a fitness landscape for a two-dimensional design space of computer processor speed and memory size. The measure of fitness is speed of performance for a given software

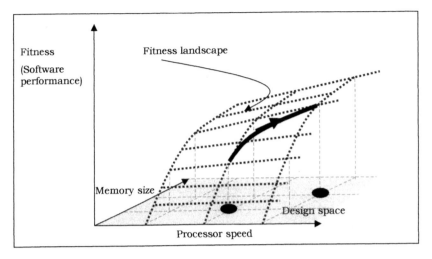

Figure 19. An example of a fitness landscape.

package. The heavy, curved line shows an increase in fitness caused by changes in both dimensions.

Change in an N-Dimensional Design Space

For our purposes the model system is considered closed, i.e., it has no external environment. I acknowledge that this is an abstraction from the real world but it is done so that we may model the system on a real-world, physical computer. The reasoning for this abstraction is that the environment outside the system we are considering has a negligible influence on the components within, and so may be ignored. When modelling a computer-based system the boundary may be drawn around the owning organisation, its customers and suppliers; the social and political environment may be ignored. On the other hand, when building socio-technical systems such as a uranium mine or nuclear reactor, the boundary includes the social and political environment.

Within the model system, components represent every part of the real system, i.e., in a computer-based information system, users are components, applications are components, communications links are components and so on, and each is part of the environment for all the others; in a society the components are the people, artefacts, resources, weather, etc., and the emergent behaviour of a society is its culture.

In the example shown in Figure 19 above, the design space consisted of two dimensions, but a computer-based, socio-technical system generally consists of a network of many co-evolutionary components, each occupying a point in a multi-dimensional design space. A change in design corresponds to a movement in the design space. This design space includes all the dimensions of all of the components, i.e., it is an N-dimensional hyperpolygon, a structure that few can visualise. In complex socio-technical systems there are very many dimensions so N is a very large number. The model described here characterises change as movement in such an N-dimensional design space.

Extending the computer system in Figure 19 to an N-dimensional design model, let us take a computer system where fitness is the speed of a particular program operation. Some possible attributes or dimensions are memory size and speed, processor speed, CD speed, hard disk speed and software program efficiency, i.e., a six-dimensional system. We are accustomed to visualising three-dimensional and even four-dimensional systems in two-dimensional pictures but beyond that most of us have problems with multi-dimensional visualisations. So in this book a convention has been adopted where each dimension is shown as orthogonal in a two-dimensional space and the value of each dimension in the component is shown as a point on the dimension line; Figure 20 illustrates the idea. The value simply indicates a normalised, possibly subjective, position in that dimension. The dimension could be magnitude, CPU speed, memory size, truth, colour, or perhaps shape. In the case of shape and colour the dimensional magnitude is clearly arbitrary. Thus, each component creates a particular path or locus in the N-dimensional space. The locus is a state description of one of the many potential designs. In the cases illustrated in Figures 20 and 21, there are 16 dimensions, and if 10 discrete values are assigned to each dimension then there are 10^{16}, i.e., 100,000,000,000,000,000 possible designs.

This locus, or state descriptor, then maps onto a single fitness point on the component's fitness landscape as shown in Figure 22. When considering change between an initial design and a modified design, the two designs map onto two points in the design space and two points on the fitness landscape. As the design changes and moves in the design space, then

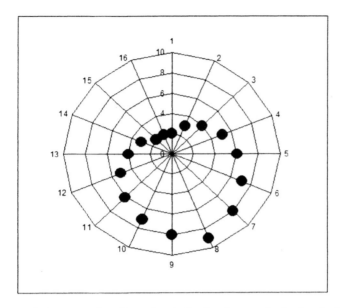

Figure 20. A design space of 16 dimensions indicating the magnitude of each of the given component's dimensions on a scale of 0 to 10.

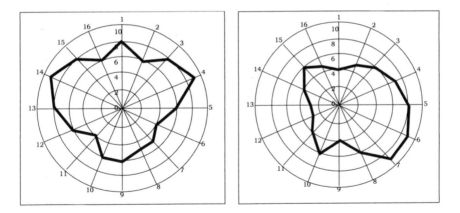

Figure 21. Two loci describing two different designs in the same design space.

potentially the fitness changes, as illustrated in Figure 23. If the fitness does not improve, it may be that the change will not occur. Alternatively, the pressure for change may be so large that the change occurs even though it causes a reduction in fitness.

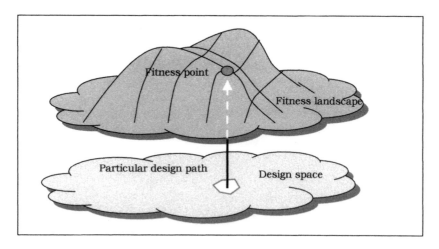

Figure 22.   Mapping the design onto the fitness landscape.

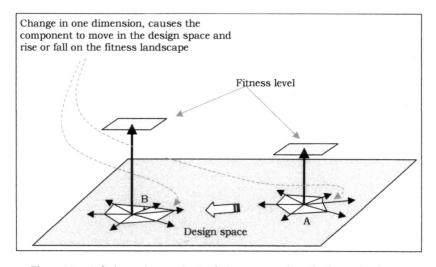

Figure 23.   A design point moving in design space and on the fitness landscape.

The Co-Evolutionary Process

A movement in the design space, i.e., a change in a component in any given time-step, depends on factors derived from influences from all the other components in the system as illustrated in Figure 24. In a modelled

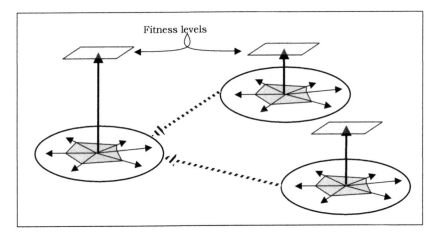

Figure 24. The fitness algorithm of each component takes into account the influence of its environment.

system, all changes happen simultaneously within the time-step. For modelling purposes, time proceeds in discrete steps but these time-steps can be arbitrarily small. At the next time-step, the changed components generate new environmental pressures within the system and the process is repeated. In other words, the process is co-evolutionary, each component influencing the others.[79,83]

Not all components have the same level of coupling and of influence on others; compare the influence of a supervisor on an employee with the influence of the employee on the supervisor. The model of the effect of the pressure one component has on another is a network of nodes linked by asymmetric connections, i.e., the directed graph we met before. The pressure for change on one component is an aggregate of the pressures from all those components on which it is dependent; the same is true for all the other components so the net effect of successive pressures and changes is co-evolution.

In the model every component is represented by a node, every node is connected to every other node, and the connections all have a coupling or influence weighting ranging from 0.0 to 1.0. When computing the fitness of a node the algorithm uses the weighting to attribute the magnitude of the effect of the influencing nodes to the pressure for change. The dimensions of the system under investigation are combined in a single line in the structure of the model.

The Threshold

The central part of a change is the decision. Within a component, whether or not a decision for change occurs is dependent on two factors, cumulative pressure for change at this time-step and the component's resistance to change. Each of the components, social and technical, may be viewed as autonomous agents with an internal goal which may be rational, irrational or programmed. The aggregation of the goal, the effects of the control of change, the cost of change, the risk of change and the environment determines the value of the resistance of change for each dimension.

A "change algorithm" determines whether the change threshold has been crossed, a "transition algorithm" then determines the magnitude and dimensional nature of the change. Of course, in the real world, there are no tangible algorithms but the effect is the same. The schematic in Figure 24 and the diagram in Figure 25 map to the following protocol:

> A change will occur if, and only if, the pressure for change is greater than the resistance to change.

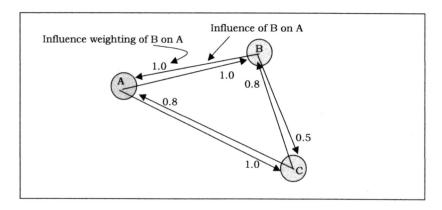

Figure 25. A structural model of the system in Figure 24.

The pressure for change is the aggregate of the goal of the component and the sum of the influences from the other components. The resistance to change is the aggregate of the risk engendered by a change to this component, the cost of a change to this component and the system level control that determines whether change will take place or not (Figure 26).

These quantities of pressure, resistance, risk, etc., have, in general, different dimensions and are largely orthogonal so the vector representation on a two-dimensional paper is a visual and mental aid instead of a truly accurate representation. Figure 27 is a high-level schematic of the change operation. This rule permits the estimation of the threshold of change, provided sufficient data is available to characterise the component factors. I will address this constraint later.

The change process in Figure 28 shows the cycle where the states of the components other than the target component generate a pressure for change on the target component. The change algorithm weighs the environmental influence and the goal of the component in terms of pressure for and resistance to change. It evaluates the effect of a change on the fitness of the component. When the component is operating under a form of control it assesses whether this control permits the change, it evaluates whether allowing the change will generate or exacerbate any risk to the component and it evaluates the cost of the change to the component. It must then perform a trade-off among all these parameters and decide if some form and magnitude of change will raise the fitness of the component. If that desirable outcome is possible then a change occurs.

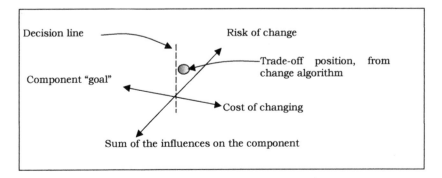

Figure 26. Diagram of the change process.

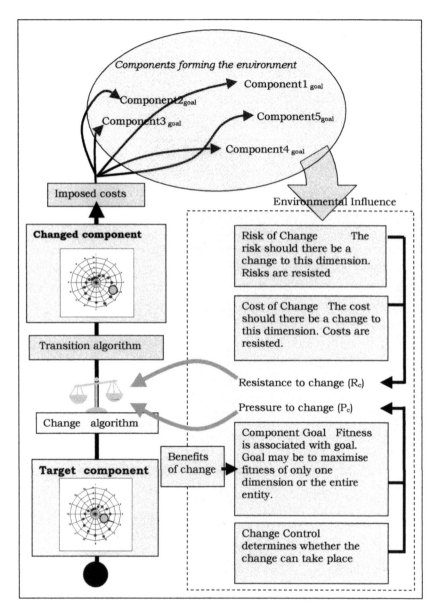

Figure 27.   Schematic of the change process.

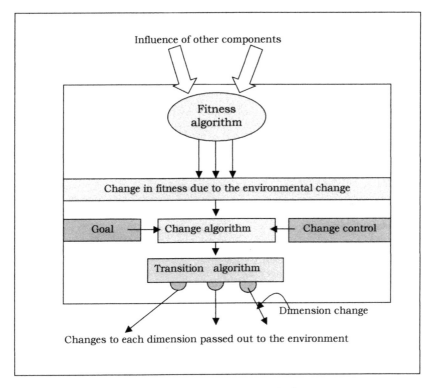

Figure 28. The change rule operating inside the component.

Figure 28 illustrates the process within the component. If the decision to change is made, the transition algorithm implements it; this algorithm operates on the change parameters given by the change algorithm. All the components pass through this process simultaneously in the interval between one time-step and the next, operating as if they were within an unchanging environment. At the next time-step, any changes become visible to the other components within the system and the external costs of the changes are applied to the components as an environmental pressure.

Resistance to Change

The statement "Change is as inexorable as time, yet nothing meets with more resistance" is attributed to Benjamin Disraeli and it seems as relevant today as in Victorian England. Resistance to change is created by the "aim" of the component. Michael Hannan and John Freeman,[84] for example, equate resistance to change in organisations to inertia, saying that "individual organisations are subject to strong inertial forces" and that these forces are dependant on the goals of the organisation with respect to their context or environment.

Resistance to change varies with the direction of the change in the N-dimensional design space and forms an N-dimensional "resistance landscape" with the component lying on the landscape. If the resistance landscape is featureless as in Figure 29, then very little pressure is needed to change the component. This may be likened to a ball on a flat surface. As the resistance landscape becomes more rugged, greater or lesser pressure is required to change in particular dimensions. The diagrams in Figures 30 and 31 illustrate this concept for a simple two-dimensional component.

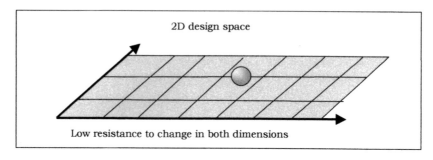

Figure 29.   Two-dimensional resistance landscape.

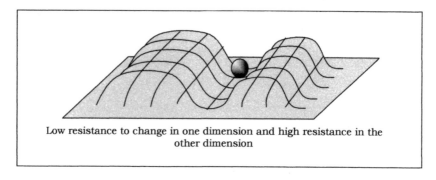

Figure 30.   Asymmetric resistance.

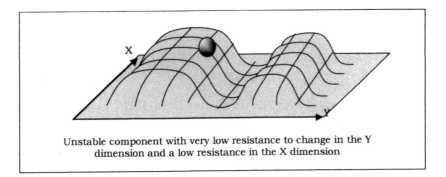

Unstable component with very low resistance to change in the Y
dimension and a low resistance in the X dimension

Figure 31. Extremely low asymmetric resistance.

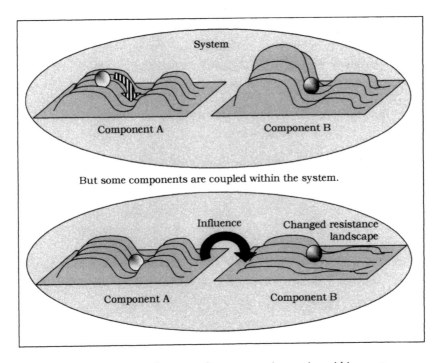

Figure 32. Resistance diagrams of components interacting within a system.

The "gorge-like" surface of Figure 30 demonstrates asymmetric resistance, i.e., the component will change fairly easily in the X dimension but will require a great deal of pressure to change in the Y dimension.

However, if the component is perched on the lip of the "gorge", as in Figure 31, it is unstable and extremely likely to change in the Y dimension but only fairly likely to change in the X dimension. Figures 30 and 31 show resistance landscapes for a single component. However, a more typical system has many components and as the components that form the environment change, so does the shape of the fitness landscape for each component.

In Figure 32, component A changes, which changes the resistance landscape for component B and hence B's options for change.

Resistance diagrams are a useful graphic aid when planning a change; they force you to identify the stakeholders and the other players (the components) in the system, to define the influences and to make a qualitative assessment of each component's resistance to moving along the many dimensions of the system. The examples above only illustrate two-dimensional resistance diagrams; one- and multi-dimensional diagrams are also useful.

6

Propagation

A change to a component is followed by the effect of that change on the other components in the environment. It is a theme of this book that the failure of socio-technical systems is most often due to the unpredictable nature of a system's emergent behaviour. This unpredictability comes from the propagation of change through co-evolving components. The effect of a change within one component is generally predictable but when many components are connected by influence links, the effects become complex. When changes propagate across a system they move the components in the design space and that movement feeds back into the succeeding changes.

The effect is not unlike the effect on a pool table when the cue ball has been struck and it changes position in two-dimensional space. The cue ball will strike one ball, then ricochet off and strike several more, each ball striking other balls in turn until all momentum is lost. Even operating under the unchanging laws of physics, the final distribution of balls on the table is emergent and depends on the skill of the player, the original distribution of the balls, the condition of the table surface, etc.

A key characteristic in the propagation of change is the interference one change has on another. Thus far, we have concentrated on the influence of the environment causing a single change in a single component. We have ignored the cumulative effect changes have on the environment and hence on subsequent changes. This interference causes the apparently indeterminate character of emergent behaviour. It is indeterminate only because it is not humanly possible to define every initial condition and the status of every component after an interfering change has taken place.

The Temporal Effect

This temporal effect is probably best illustrated by the diagrams in Figures 33 and 34. In these diagrams the circle representing a component

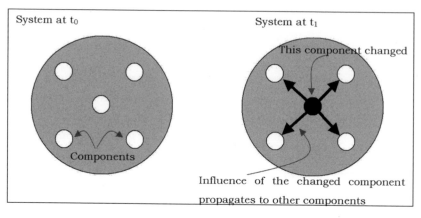

Figure 33.   First stages of propagation.

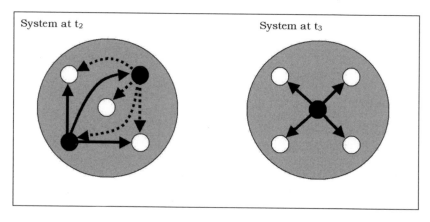

Figure 34.   Next stages of propagation.

is coloured black to indicate that the component has changed between the last time-step and this time-step. The black arrows indicate that the change in the component at the base of the arrow influenced the component at the tip of the arrow.

Each component evaluates its internal state and external environment and "decides" whether to make a change using its threshold rule. At time-step t_1, one component has changed, influencing the other components to change. At time-step t_2, the change in one component has caused two of the components to change. All the other components now have two influences pushing them to change. At t_3, the influence for change has caused the original component to change again, giving feedback to the original change. This causes further but possibly different changes to propagate.

An Illustration of Propagation in a Model

A simple way is see what is going on is to reduce the complexities of the real world to a model. To start with, each component is restricted to a single attribute with two states — extrapolation to many attributes can come after I discuss the simple case. It is also assumed that the behaviour of a component to a change in its environment is deterministic, i.e., follows a known rule. Both are sweeping simplifications but are sufficient to illustrate the magnitude of the problem. Figure 35 shows such a system.

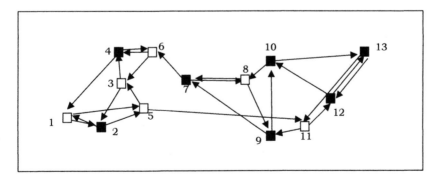

Figure 35. A system whose components have a single attribute and two environmental influences.

In general, if S is a state selected from n states, E is the set of influencing components, i.e., the component's environment, and I is the deterministic rule for change, then the general rule is:

$$S(t+1) = I(St, SEt).$$

Of the modelling systems available, I chose one-dimensional cellular automata as the closest model to the system in Figure 35 and that giving the most graphic state/time description of the system dynamics.

A cellular automaton consists of a number of cells, each with the ability to exist in one or more states and organised in a regular lattice of one or more dimensions. The state of each cell at time $t+1$ depends on a single set of rules that take into account the state of the cell at time t and its environment consisting of its neighbouring cells at time t. This simple construction can exhibit a global emergent behaviour not present in the local rules. Cellular automata may be envisaged as a system consisting of a number of identical components with identical rules, hence the system's transition from one time-step to the next is deterministic.

Their advantages and disadvantages as a model are discussed in Chapter 8 but their major advantage is that any system with many identical discrete elements undergoing deterministic local interactions may be modelled as a cellular automaton. The example system shown in Figure 35 is shown again in Figure 36, rearranged to clarify the mapping to one-dimensional cellular automata.

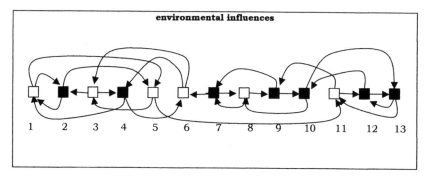

Figure 36. The same system rearranged.

In cellular automata, the "cells" are so called from their origin in the field of self-reproducing machines[85,86] and their states can range from two to many. In this case, the components have been restricted to two states and an environment of two neighbours. In the diagrams below, the entire row of cells represents the state of the whole system at the current time-step and the colour of each cell represents the state of the individual component. The set of states within the system is derived from the previous time-step by applying a group of simple rules; the same group of rules is applied to each cell in the row.

The change rules, in this example, become:

1. Examine a designated cell and the cells on either side of it, i.e., its environment (Figure 37).
2. Determine which of the upper patterns the set of three cells conforms to in the rule set (Figure 38).
3. Colour the cell below the designated cell the colour of the single cell below the selected pattern. The colour represents the cell state (Figure 39).

In the simple system the colour of the cells in each successive row is changed to that determined by the pressure of the designated cell's

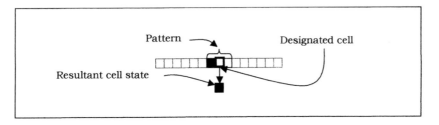

Figure 37.   A cellular automaton's change rule.

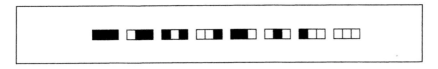

Figure 38.   The eight possible permutations of the three states.

environment and the rules in the transition algorithm. In this case, it is assumed that the system always bows to environmental pressure and accepts the change. In a one-dimensional cellular automaton each target cell and its immediate neighbours form a set of three and each cell in the set has a choice of two states, hence there are 2^3 or eight possible permutations, as in Figure 38.

Each of the eight sets in Figure 38 can give rise to a target that is either black or white, as shown in Figure 39. This amounts to a rule set governing the changes.

There are eight cells, each with the possibility of two states, so there are 256 possible outcomes (00000000 through to 11111111) associated with the permutations of target states in Figure 38. What happens if we set up a row of cellular automaton with a set of rules as in Figure 39 and set the initial states as shown in Figure 40?

As we can see from Figure 40, most of the cells are zero, as are their neighbours, so they remain zero. However, there are three sets of cells that

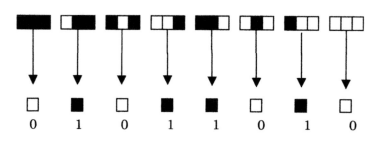

Figure 39.   The graphic rule set for the target cell.

Figure 40.   A cellular automata at time t_0.

Figure 41. A cellular automata at time t_1.

are not all zero. These are the central sets. As before, we will assign 0 to white and 1 to black so we can say they are 001, 010 and 100. The 001 set changes its central cell from 0 to 1, the 010 set changes its central cell from 1 to 0, and the 100 set changes its central cell from 0 to 1. At time t_1, after these changes, Figure 41 shows the system state.

If we show the successive system states in one diagram we can see how the system state changes through time, remembering that all the cells are using identical transition rules.

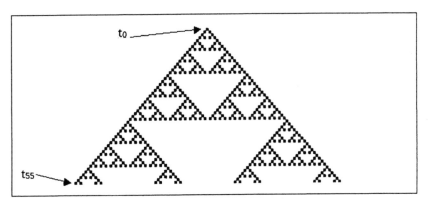

Figure 42. Successive rows of system states; the borders of the white or 0 cells are not showing.

Figure 42 shows how the changing of the state of one cell from 0 to 1 propagates to the limit of the system over time. If we examine the system state at any of the later time-steps, it is unlikely that we could have forecast it without going through the previous steps. Even with this extremely simple system, forecasting any future state after making a single change is a very uncertain process.

7

Modelling and Modelling Mechanisms

It is clear that the system's behaviour over time is the critical factor in determining whether it will satisfy the expectations of its sponsors. Hence, before going on to describe the modelling of a complex system's behaviour, it is necessary to discuss the technical aspects of models of socio-technical systems with a view to determine the necessary attributes of a mechanism for such models. This chapter also discusses the soundness of using computer programs to simulate model behaviour, and looks at a number of mechanisms that have the potential to form the basis of a computer-simulated model.

Models

"With a good model comes discovery, with discovery comes understanding, with understanding comes control."[87]

As we have seen, manual analysis of any reasonably sized, complex system would take a considerable time; and may indeed be as intractable as the three-body problem. The emergent behaviour of a complex, co-evolutionary system is therefore a candidate for modelling and computer simulation. I will outline the advantages of modelling and explore the use of computer simulation to provide insights into the process of change; we saw its usefulness in visualising propagation with cellular automata at the end of the last chapter. Building a model has long been recognised as a way of understanding the world; something that everyone does but which science and mathematics has formalised. A model is a simplification of a concept, structure, or system. A toy car or even a drawing of a car is recognisably a car, even if it is much smaller and less complex than a real car. Nevertheless, car manufacturers build a range of models before building real cars, to test performance, wind resistance, and other attributes in a more cost-effective fashion. Simulation of models of socio-technical systems is equally cost-effective.

Computer-Based Models

Nigel Gilbert and Klaus Troitzsch discussed the predictive use of simulation, saying,

> "If we can develop a model which faithfully reproduces the dynamics of some behaviour, we can then simulate the passing of time and thus use the model to 'look into the future'. A relatively well-known example is the use of simulation in demographic research, where one wants to know how the size and age structure of a country's population will change over the next few years or decades. A model incorporating age specific fertility and mortality rates can be used to predict population changes a decade into the future with fair accuracy."[88]

This predictive use of simulation can be usefully extended to socio-technical systems, and Robert Axelrod and Michael Cohen declare that "Exploring an issue via simulation is a viable scientific approach when the problem at hand is sufficiently complex to defy more traditional alternatives."[40]

Data collection and simulation via stochastic methods — that is, by initialising the system with randomly generated variables and processing it through multiple time-steps — allows us to better understand the underlying behaviour. The emphasis is on simulation because real-life socio-technical systems and other wicked problems are too complex to undertake such analysis by hand. Further, Ross Ashby reminds us that stochastic methods should be used where the system at hand is both large and complex.[19] The abstraction of simulation loses many of the attributes of the real system but it does raise the possibility of insight that would otherwise remain hidden.[89]

The major advantages of computer-based models are:

- they are explicit, i.e., their assumptions are stated in the written documentation and are open to all for review;
- they infallibly compute the logical consequences of the modeller's assumptions;
- they are comprehensive and able to interrelate many factors simultaneously.[90]

If they infallibly compute the logical consequences of the modeller's assumptions, where can new knowledge come from? Herbert Simon's replied,

"There are two related ways in which simulation can provide new knowledge — one of them obvious, the other perhaps a bit subtle. The obvious point is that, even when we have correct premises, it may be very difficult to discover what they imply. All correct reasoning is a grand system of tautologies, but only God can make direct use of that fact. The rest of us must painstakingly and fallibly tease out the consequences of our assumptions. ... The more interesting and subtle question is whether simulation can be of any help to us when we do not know very much initially about the natural laws that govern the behavior of the inner system"[5] pp. 14–15.

Both of Simon's "ways" may be applied to the study of change in complex, co-evolutionary, socio-technical systems. First, because of the interdependence between the multiple components, the only practical way to test assumptions about the properties of change in such systems is by simulation and, second, because it may illuminate the "inner laws" of system change that still remain to be discovered. Empirical, real-life investigations could be made by monitoring and controlling, for example, the building of a large computer-based system, but the cost of running multiple parallel streams of implementations renders this impractical.

Ned Gardiner[91] reinforces Simon's view, commenting, "simulation offers insights into patterns and behaviour which would be experimentally elusive". He made this comment in the context of ecology in his introduction to The Santa Fe Institute's Complex Systems Summer School — The Long Term Ecological Research Network. He added,

> "Whether your research is purely theoretical or entirely empirical, modeling can be used (1) to synthesize knowledge and (2) to compare expectations, simulation results, and observable phenomena. The necessary investment of effort by ecologists is simple to justify: simulation offers insights into patterns and behavior which would be experimentally elusive, at the very least. Further, modeling is the principal avenue by which ecologists may draw from parallel research in complex systems by physicists, economists, and computer scientists."

Thus, in addition to the empirical evidence that a multitude of simulations are being used to practical effect in a large variety of disciplines; the scientific literature supports the idea that dynamic simulation of a model can give insights into the behaviour of complex, socio-technical systems under change.

Aspects of Model Design

The vast majority of socio-technical systems break down into a number of components with well-defined boundaries, e.g., sponsors, users, software, communications, etc., as in Figure 43. When these components are arranged in a hierarchy, with each major system breaking down into a number of medium-sized systems that in turn break down into yet smaller systems, it is called a "system of systems".[92,93] Whilst the phrase "system of systems" does not advance our understanding beyond Simon's descriptive "hierarchy", Mark Maier[94] describes five principal characteristics that in his view are "useful in distinguishing very large and complex but monolithic systems from true systems-of-systems". Maier's "systems of systems" have the following characteristics:

1. Operational Independence of the Elements: If the system-of-systems is disassembled into its component systems the component systems

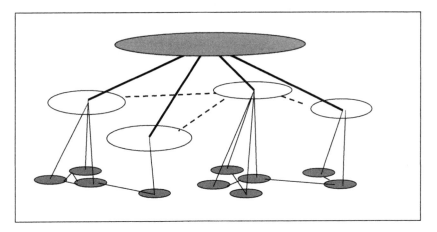

Figure 43. A system of systems.

must be able to usefully operate independently. The system-of-systems is composed of systems that are independent and useful in their own right.

2. Managerial Independence of the Elements: The component systems not only can operate independently, they do operate independently. The component systems are separately acquired and integrated but maintain a continuing operational existence independent of the system-of-systems.

3. Evolutionary Development: The system-of-systems does not appear fully formed. Its development and existence is evolutionary with functions and purposes added, removed, and modified with experience.

4. Emergent Behaviour: The system performs functions and carries out purposes that do not reside in any component system. These behaviours are emergent properties of the entire system-of-systems and cannot be localised to any component system. The principal purposes of the systems-of-systems are fulfilled by these behaviours.

5. Geographic Distribution: The geographic extent of the component systems is large. Large is a nebulous and relative concept as communication capabilities increase, but at a minimum it means that the components can readily exchange only information and not substantial quantities of mass or energy.

Whether we call the system under consideration a complex system or a system of systems, the simulation mechanism needs to be able to map its structure onto the model via a mechanism of weighted influence connections.

The influence for change can flow up and down the connections between the different levels of components and between the components. The granularity of the system is determined by the view taken of it. At a broad level, the granularity is large, the number of granules small, but at a detailed level the granularity is small, and the number of granules is large. Additionally, the granularity of each dimension of each component may be coarse or fine. Time may be modelled by decade, year, day or second. Nation states, international companies, small businesses or individuals may be used to model an organisation. Choosing the dimensions and their granularity is one of the most important tasks in building a model. If the granularity is too coarse, the model will be too inaccurate to be useful. If it is too fine, the tasks of model building, data collection and computation will be unachievable. The rate of change is also largely dependent on the size of the granularity and the layer of the system in which it exists; in general, the larger the granule and the higher the layer, the slower the rate of change.

Coupling

When we talk about one component influencing another, we say one component is coupled to the other. There are two forms of coupling in a complex system, that within a component (software designers call this cohesion[95]) and that between components. In this book, I refer to them as intra-component coupling and inter-component coupling, respectively. In both forms of coupling there is a range of degree, going from highly coupled to not coupled at all. However, the meaning differs between the two forms and the simulation needs to be able to map both forms into the model.

Intra-component coupling

High intra-component coupling is defined as a strong interaction between the design space dimensions of the component, and a change within one dimension can cause a simultaneous change in the other dimensions of the component. Hence, achieving a higher fitness can be subject to heavy trade-offs.

Taking a personal computer as an example of a component, the price and CPU speed are strongly coupled and changing one will probably change the other. Alternatively, low coupling means the dimensions of the component are largely independent and a change in one dimension has little effect on the others. Again, taking the personal computer as an example, the colour of the cabinet and the CPU speed are not coupled at all.

I define the degree of intra-component coupling as the aggregate of the binary relationships, the relationships being defined as interacting or non-interacting, one or zero. In an N-dimensional component, each dimension may interact with each of the other $N-1$ dimensions. The total number of links is $N-1+N-2+\cdots+N-N$ or

$$N^2 - \sum_{i=1}^{N}(i) \tag{1}$$

where the group of symbols $\sum_{i=1}^{N}(i)$ means the sum of all the numbers from 1 to N.

If we assume, as we have done before, that the interactions are uni-directional, then the number of potential interactions is twice the number of links, i.e., one in each direction.

$$2 * \left[N^2 - \sum_{i=1}^{N}(i) \right]. \tag{2}$$

The number of actual interactions may be fewer as not all components influence those that influence them. I will denote the actual interactions as A. Hence, the degree of intra-component coupling is defined as:

$$\frac{\left[2 * \left[A^2 - \sum_{i=1}^{A}(i) \right] \right]}{\left[2 * \left[N^2 - \sum_{i=1}^{N}(i) \right] \right]}. \tag{3}$$

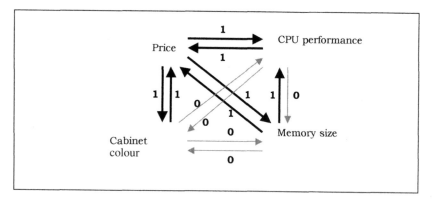

Figure 44.   Intra-component coupling.

For example, in a component of four dimensions, and continuing with our personal computer example, the dimensions are price, CPU performance, memory size and cabinet colour.

As indicated in Figure 44 there are 12 potential interactions but only 7 actual interactions.

- Price influences CPU performance, memory size and cabinet colour,
- CPU performance influences price,
- memory size influences CPU performance and price, and
- colour influences only price.

Thus the degree of intra-component coupling (D_{icc}) is

$$D_{icc} = \frac{2*(49-(6+5+4+3+2+1))}{2*(144-(11+10+9+8+7+6+5+4+3+2+1))}$$

$$= \frac{28}{78} = 0.34.$$

So D_{icc} is an indication of how closely entangled the component's dimensions are, and is a measure of how likely one dimension will be affected by change in another dimension.

Inter-component coupling

A high degree of inter-component coupling means the component boundaries are weak and changes in the surrounding components are likely to cause changes within the coupled component. Typical examples are predator/prey systems. High inter-component coupling can cause cyclic propagation of change, i.e., feedback to the component that changed first, perhaps changing that again and causing further cycles of change; for example, the "howling" heard within sound systems when the coupling between the loudspeakers and microphone is too high.

On the other hand, a low degree of coupling means strong component boundaries and such components are largely impervious to changes in the environment. This situation generally occurs only in the higher layers of complex systems, but when change does occur it is often catastrophic, for example, where the significance of a change in procedure in one airline department is not recognised. Inter-system coupling of this change to other departments can cause unexpected, disastrous effects such as an aeroplane crash.[96] Another example is where an apparent change for the good in the Okavango delta led to a change from subsistence farming to starvation.[4] The coupling between two components is uni-directional. That means a change in component A can cause a change in component B, but the process is unrelated to whether a change in B can cause a change in A. The complexity of an inter-component coupling lies in the magnitude of the influence of each strand on the influenced component. Figure 45 illustrates this idea.

Inter-component coupling can be defined as the average of the magnitude of influence of the strands:

$$\frac{\sum_{n=1}^{N} M_n}{N}$$

where N is the number of attributes and M is the magnitude of the influence. I will explain this with an example. Suppose, with six dimensions in the influencing component, there are six strands with the influence magnitudes of 0.1, 0.3, 0.1, 0.8, 0.5, 0.0 on the target component. Then the

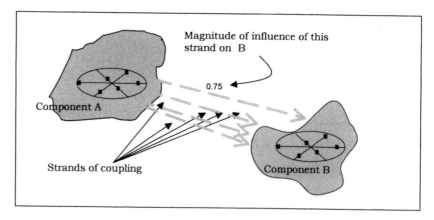

Figure 45. Strands of coupling.

degree of inter-component coupling is

$$\frac{(0.1+0.3+0.1+0.8+0.5+0.0)}{6} = \frac{1.8}{6} = 0.3.$$

This number gives an indication of the strength of the coupling between two components, and if the couplings between all the components are averaged and the standard deviation taken, then the system's overall coupling can be characterised.

Cumulative Effect

When a strong environmental pressure is applied to a component, it is easy to imagine the component's threshold for change being exceeded and a change taking place. On the other hand, the model also needs to take into account the cumulative effect of many small environmental pressures. These small effects can be illustrated by two examples.

1. Many small changes that finally reach the threshold for a major, perhaps catastrophic, change. For example, the many instances where slow accumulation of snow on a roof causes it to collapse.[97]
2. Many environmental components, each exerting a small pressure for change but collectively providing sufficient pressure to change the

target component. Page and Shapiro[98] mention the example of the pressure of public opinion producing a change in national policy. The political pressure of one ordinary individual is negligible but when many express the same opinion, a government is pressured to change its policy if it wishes to stay in power.

Structural Change

The propagation of change depends on the structure of influence connections. The organisational structure plays a major role in the behaviour of the system as indicated in Chapter 4, and as time passes, influence connections may be formed, strengthened, weakened and destroyed. An employee may move from one company to another and thus be subject to different influences and influence different people in turn, or a software component that grows in utility has a greater influence on the way people work. Therefore, in addition to accommodating the dynamics of component change propagation, the simulation needs to be able to modify the influence connections, both manually and dynamically, by means of rules.

Mechanisms

I have looked at a number of mechanisms with which models might be created to simulate complex, co-evolutionary systems and to provide insight into the properties of change. Two of these mechanisms are described below.

Cellular Automata

You will recall that a simple construction using a cellular automaton can exhibit a global emergent behaviour not present in the local rules. The change rule is local in space and time but the total effect of all the local changes is a global behaviour pattern. As we saw before, a global pattern emerges even if the initial state of the cells is selected randomly as in Figure 46.

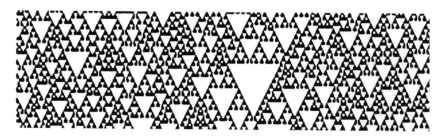

Figure 46. A state/time diagram of a one-dimensional, two-state cellular automata, one horizontal line per time-step.

Conway devised a set of very simple rules for a two-dimensional, two-state cellular automata that have enormous computational power.
His rules were:

- Assign 0 or 1 (shown as white and black, respectively) to the two states;
- At each time-step, sum the states of the eight neighbours around the designated cell;
- If the sum = 2 then the cell state does not change;
- If the sum = 3 then the cell state becomes 1;
- If the sum is any other number then the cell state becomes 0.[99–101]

He called the automata governed by these rules the "Game of Life" because of the variety of patterns that emerged. The state/time diagrams of the Game of Life as shown in Figure 47 illustrate the development of the Glider Gun pattern. It is called the Glider Gun because it gives rise to a set of periodic small patterns (gliders) that move across the lattice; it cycles through 30 steps before repeating the pattern and generating another glider. It is important to note that only the emergent pattern moves across the lattice. Individual cells remain motionless, simply changing state as local rules dictate.

Cellular automata were introduced originally by John von Neumann and Stanislaw Ulam as simple models for studying self-reproduction,[85,86] Stephen Wolfram took the next step and analysed the behaviour of cellular automata considered as systems.[102,103]

Wolfram has spent many years studying cellular automata and has published a book *A New Kind of Science*[104] which claims that the natural world runs on simple programs such as cellular automata. In this, he

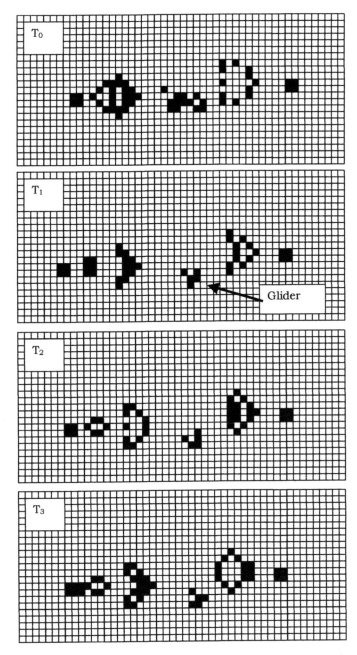

Figure 47. The Game of Life. A state/time diagram of a two-dimensional, two-state cellular automata running from time T_0 to T_3, one lattice per time-step.

followed Konrad Zuse who in 1967 even suggested that the entire universe is being computed on a cellular automaton.[105,106] Wolfram writes,

> "Three centuries ago science was transformed by the dramatic new idea that rules based on mathematical equations could be used to describe the natural world. My purpose in this book is ... to introduce a new kind of science that is based on the much more general types of rules that can be embodied in simple computer programs. ... the intuition [is] that creating complexity is somehow difficult, and requires rules and plans that are themselves complex. But the pivotal discovery that I made some eighteen years ago is that in the world of programs such intuition is not even close to correct. ... some of the very simplest programs that I looked at had behavior that was as complex as anything I had seen. ... it implies a radical rethinking of how processes in nature and elsewhere work. ... I have been led to a ... sweeping conclusion, summarized in what I call the Principle of Computational Equivalence: that whenever one sees behavior that is not obviously simple — in essentially any system — it can be thought of as corresponding to a computation of equivalent sophistication.
>
> ... But from the Principle of Computational Equivalence there also emerges a new kind of unity: for across a vast range of systems, from simple programs to brains to our whole universe, the principle implies that there is a basic equivalence that makes the same fundamental phenomena occur."[104]

Wolfram's conjecture is partly based on the remarkable similarity between natural patterns and those generated by cellular automata. A typical example is shown in Figure 48; a shell showing patterning possibly derived from natural one-dimensional cellular automata, and a state/time diagram of one-dimensional cellular automata starting with randomly chosen states.

As we saw earlier, any system with many identical discrete elements undergoing deterministic local interactions may be modelled as a cellular automaton. When the effect of the local rules is non-linear, then the system becomes complex and non-trivial outputs are produced. Real-world examples can be found in patterns and forms of biological organisms,[107] snowflake growth[108] and social behaviour.[109]

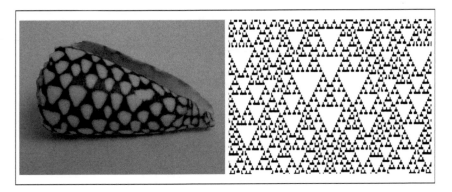

Figure 48. Shell patterning and cellular automata state/time diagram.

Conway's Game of Life[99] described above is an example of a cellular automaton that is sufficiently complex to support a pattern that behaves as a Turing machine and hence, given suitable input, is able to compute anything a digital computer can compute.[110] Therefore cellular automata may be considered as (parallel-processing) computers, in which the initial configuration encodes the program and input data, and iteration yields the final output.[69,85,102]

We saw that propagating changes can be demonstrated quite simply in one-dimensional cellular automata. The rule-based nature of cellular automata permits the incorporation of arbitrary knowledge in the transition from one time-step to the next, and are therefore useful in modelling systems over time. Examples of the use of cellular automata in simulation are described in published papers on the co-evolution of crown-of-thorns starfish and coral reefs,[111] forest regeneration[112] and epidemiology.[113]

When studying one-dimensional, two-state automata such as those above, Stephen Wolfram noted that the state/time patterns could be characterised empirically. He classified all the one-dimensional, two-state automata into one of four classes:

Class 1 — evolution leads to a homogeneous state;
Class 2 — evolution leads to a set of stable or periodic structures that are separated and simple;

Class 3 — evolution leads to a chaotic pattern;
Class 4 — evolution leads to complex structures, sometimes long-lived.

He then categorised them by turning the 256 sets of resultant states into binary numbers and hence decimal numbers, i.e., Wolfram Number 0 through to Wolfram Number 255. I will refer to "Wolfram Number" as "WN".

The set of rules illustrated in Figure 39 in the previous chapter gives the output 01011010, which is 90 when changed from binary to decimal notation, so the rule set that consists of this set of permutations and their associated resultant states is categorised as WN 90.

I believe that by applying Wolfram's claim to socio-technical systems, it may be possible to gain some insight into the workings of this subset of the natural world. First, it is necessary to decide on a suitable cellular automaton to model a socio-technical system. The model system needs to emulate a system behaviour that is regular and deterministic, so one of the Class 2 automata should be suitable. My experiments showed that there are 22 Class 2 automata that co-evolve to a regular, fractal state/time form when seeded with a single black or white cell. This group of automata all display the regular behaviour expected of a simple co-evolutionary system. The behaviour of the WN 151 rule set is typical of the group so I chose it, arbitrarily, as the representative rule set.

I wrote a simple program to reveal the evolution of any rule set under various seed conditions. Figure 49 shows a screenshot of this program after the two-neighbour WN 151 rule set had run for multiple time-steps, starting from a single white seed cell. The program has the capacity to also evolve four-neighbour rule sets but detailed investigation of the behaviour of the 2^{32}, i.e., approximately 4,000,000,000, four-neighbour rule sets has been left for another book.

Simultaneous Change

Using the automata with WN 151 to generate stable, periodic patterns over time, it is possible to use this change rule to represent the regular operation of a system from one time-step to the next. The change rules for WN

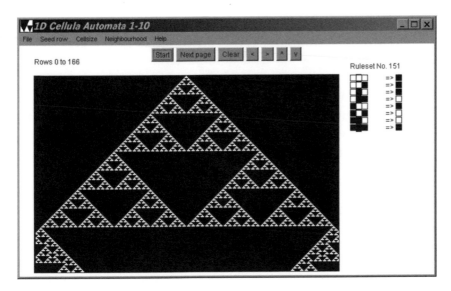

Figure 49. The one-dimensional CA program showing the evolution of the WN 151 rule set.

Figure 50. Time-step T_1.

151 are most easily shown graphically as in Figure 49. The simulation starts with a row of cells, all of which are in the "black" state, and the WN 151 rules are applied. It is clear from Figure 49 that the only rule invoked is the left-most rule and hence there is no change over time. If at time-step T_1 the state of one cell is changed to white (Figure 50) and the rules are executed successively at each time-step, we get a regular changing pattern over time, as shown in Figure 49. From time-step to time-step, the pattern of states is predictable, just like in a simple physical system such as a pendulum.

Whilst exploring the behaviour of cellular automata I noticed that, within a single system, when one time series of changes intersects with another, they cause interference in a manner reminiscent of the interference of light or sound waves. This interference due to two propagating time series intersecting was used to illustrate the situations that occur when the effects of one change interferes with the effects of another change. For the next experiment, we start with a black row but change two cells to white a given distance apart as in Figure 51, representing two components or changes introduced into a system at the same time.

At first, there is no interaction but eventually the two propagations of state changes interfere and a new pattern appears. The simulation shows that different patterns are generated depending on the distance between the original two white cells and hence the point in the cycle at which they interfere. In Figure 52, the left-hand state/time diagram

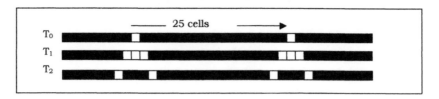

Figure 51.   Two changes 25 cells apart, shown after 3 time-steps.

Figure 52. Interference in cellular automata.

shows that the pattern changes when the effects of the two original changes meet, but it is still regular and predictable. However, in the right-hand diagram, a very different pattern emerges when the two sets of effects meet.

When such widely different behaviours are initiated by a very small change many time-steps before the interference occurs, it seems quite plausible that designers of complex systems would be unable to predict if their system will behave like A or B after a few changes.

The only difference between these two pictures is that in B the initial two white squares are one position further apart!

Subsequent Changes

After further simulations, it became apparent that when changes occur at different times, the interference could lead to deterministic chaotic behaviour where it is virtually impossible to foresee the changes.

In this context, deterministic chaos is indicated by every succeeding set of states being different. This behaviour is demonstrated in Figure 54.

In both screenshots the WN 151 rule set was invoked for every cell at each time-step, i.e., in each row. In the example on the left, the initial row consists of black cells with two white cells, 25 cells apart. In the other example on the right, the initial row consists of black cells with one white cell. After a few iterations of the set of rules, one of the black cells is changed to white in the "column" 25 cells right of the initial white cell as in Figure 53. Both automata were iterated about 60 times.

It is clear that in both cases interference takes place. In the left-hand example the resultant pattern is regular but in the right-hand example, in

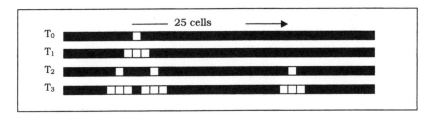

Figure 53. Second change introduced at the third time-step.

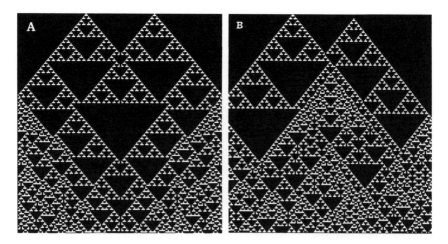

Figure 54.   Example of chaotic behaviour induced by a change later in time.

this set of circumstances, once the two sets of propagating states meet, the pattern becomes completely chaotic. The pattern is no longer regular or cyclic. Each succeeding set of states is different from the preceding set. Of course, every step is deterministic and can still be predicted from the last set of states. Yet, at the time when the change is introduced, predicting the state of the system at an arbitrary time in the future is an extremely uncertain process. This phenomenon again illustrates the uncertain result when a complex system is changed. Investigation using this cellular automaton has shown that introducing a second change before the initial change has reached equilibrium results in three different types of state/time pattern. I have characterised these propagation patterns as:

- *Regular:* the change has no effect on the regular propagating pattern;
- *Semi-Regular:* the change alters the pattern but the resulting pattern has a regular form (Figure 54A); and
- *Chaotic:* the change causes the pattern to become chaotic (Figure 54B).

Additional simulations with the WN 151 rules have shown that if the change is introduced within the set of propagating changes, the proportion

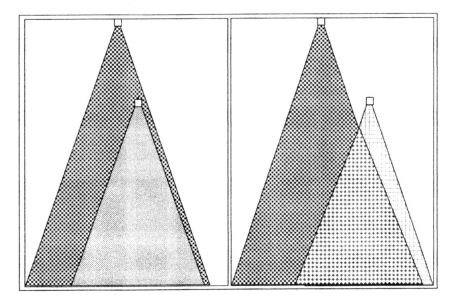

Figure 55. Change inside and outside the current propagating changes.

Table 1. Regular, semi-regular and chaotic change.

	Inside the set (%)	Outside the set (%)
Regular	39	2.5
Semi-Regular	17	43.5
Chaotic	44	54

of each propagating pattern type is different from that when the change is introduced outside the set.

Figure 55 clarifies this idea and the results are shown in Table 1.

These numbers indicate that a second change within the set of components affected by the first change has approximately 0.44 probability of causing the system's emergent behaviour to become deterministic chaos. In other words, such a change has a good chance of making the system unpredictable. If the change occurs outside the set of components affected by the first change, the effect is delayed until the two propagating sets of changes interfere, but when they do the probability of causing the system's emergent behaviour to become chaotic is even higher at 0.54. Table 1 indicates that there is approximately one chance in two of the system becoming unpredictable when a second change is made while the first change is still affecting the system. It is interesting to note that this bears out the common saying that "Disasters occur when two things go wrong at once".

Feedback

Do not forget that these diagrams are temporal, not spatial, diagrams with each *row* representing the state of the complete system at a given time. Thus, they also show the effects of feedback across the system. The changes to the cells on either side are influenced by the state of the central cell and at the next time-step these changes feed back as influences on the central cell. The same is true right through the system; over time the changes propagate across the row of cells, and are fed back.

The system illustrated in the cellular automata model and Figure 56 is similar to that in Figure 35, but is closer to reality in that the links between the components range from none to many. This alters the emergent system behaviour from that of the cellular automata model and introduces the possibility of cyclic feedback. This feedback can exacerbate the uncertainty but may also damp out the tendency to chaos.

Another point to note is that while the argument above utilised deterministic rules, the systems of interest to us are socio-technical systems that

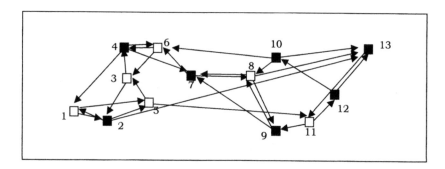

Figure 56. System with a variety of links.

include human actors whose actions, in the context of the system, are often non-deterministic. Human actors are influenced by personal assumptions and agendas, and by circumstances that would appear to lie outside the boundaries of the system. Occasionally they act irrationally, making it extremely difficult to foresee the effect of changing one variable in a complex, socio-technical system. As noted above, where road safety lobbying prompts the introduction of traffic lights on a main road, the concomitant delays can cause drivers to decide to divert down suburban streets, causing potentially greater damage and personal harm.

In any complex, socio-technical system, making two or more technical changes almost simultaneously has a good probability of causing a system to become unpredictable; the unpredictable behaviour of the human actors exacerbate this effect. Implementing a large technical system is a complex undertaking in itself and the introduction of human "components" raises the level of complexity. The biggest problem within system design and implementation is predicting the effect of the introduction of a component, an architectural change, or a change within a component. Often the change is deliberate and is designed to achieve a particular effect, and the designer and implementer are bewildered when the change achieves a different effect. Probably the most well-known example among computer system builders is making a change to one line in a program. In the context of the current module, the change is syntactically and semantically correct, but the change causes the system to crash because of a subtle interaction with another module. Is that poor design or poor implementation or a lack of holistic understanding of the system?

Although they did give us insights into the nature of change, cellular automata do not provide a mechanism for modelling influence weighting or differing transition periods for a change, so I will now look at agent-based models.

Agent-Based Models

The term agent-based modelling refers to a collection of computational techniques in which individual agents and their interactions are explicitly simulated, and emergent behaviours observed.[33] These agents are "individuals", in that their external behaviour is governed by an internal transition function that can be unique to that agent. Luis Rocha usefully defines an agent as an entity that has some autonomy of action and he comments that "This leads us to an important definition of an agent from the XIII century, due to Thomas Aquinas: an entity capable of election, or choice."[114]

If we model a system component as an agent, then the agent's transition function determines how it reacts to its internal (design dimensions and fitness) and external (environment) states to produce a new set of dimensional values, connections and influences. In the model, a transition function can be either a set of non-linear equations, a set of rules or a combination of the two. An unchanging transition function can provide complex behaviour but if the agent's transition function is mutable, the agent can also exhibit adaptable behaviour.

Davidson characterises an agent's attributes as:

- Pro-activeness, ranging from purely reactive entities to pro-active fully autonomous entities,
- communication language, ranging from having no communication at all between entities, via simple signals, e.g., procedure calls, to full agent communication languages,
- spatial explicitness, ranging from having no notion of space at all, to letting each entity be assigned a location in the simulated physical geometrical space,
- mobility, ranging from all entities being stationary to each entity being able to move around in the simulated physical space,
- adaptivity, ranging from completely static entities to entities that learn autonomously, and modelling concepts, ranging from using only traditional modelling concepts to using mentalistic concepts, such as beliefs, desires, and intentions.[115]

Agent-based models, like cellular automata, are discrete event simulations. There are two forms, time-driven and event-driven.

- In the time-driven model the simulated time moves in constant time-steps and the interaction between the agents is evaluated at every step.
- In an event-driven model the next time-step occurs when the next event occurs; the occurrence of events is driven by a time-stamped

event list. The simulation takes the first event from this list, sets the simulated time to that of the time stamp of the event, and simulates the effect on the system state. The event-driven model is more efficient because steps where there are no changes take zero time. On the other hand, it does require that a set of events are devised beforehand to impact on the simulation.

Where the effects of change in socio-technical systems are being investigated, a hybrid system is required. It is necessary to use a time-driven model to simulate the effects of change internal to the system. It is also necessary to inject changes from time to time to test the effect of a specific change.

Agent-based models are intended to show the interactive behaviour of multiple agents, and as such are themselves examples of co-evolutionary systems. Therefore, they are an ideal mechanism to simulate socio-technical systems. They have been used to simulate many individual-based systems as varied as the flocking and schooling behaviour of birds and fish,[116] artificial chemistry,[117] the human immune system,[118] and modelling civil violence.[119]

Each agent needs to be aware or be able to "sense" the state of all the other agents to which it is coupled. In the model, we know that each agent or component has a set of attributes that can change state from one time-step to the next. So, if an agent changes, an "advertisement" of the new state of the attributes is made available to the system. Each agent can look at the "advertisements" of individual or groups of agents that are coupled to it and input the contents of the "advertisement" to its own transition function. The behaviour of an agent-based model derives from the cumulative behaviour of the individual agents and is frequently counter-intuitive. This was particularly so in the behaviour of the Team simulation described in the next chapter.

Specific advantages of agent-based modelling are:

- the close match between real-world entities and the agents modelling them, as the agent's attributes can represent physical values (such as weight or colour) and abstract values (such as emotions);
- that the structure of the simulated system can be made to match the real structure; and,
- the discrete agents are discrete, which may be introduced into and removed from the simulation as it proceeds.

8

Simulation

To experiment with agent-based simulations, I wrote a program to test the premise that a genetic algorithm can be used to develop and selectively improve cooperative behaviour between agents (the meaning of the term genetic algorithm will become clear later). The simulation took the form of teams of five agents that competed in a series of tournaments; being unimaginative, I called the program "Team".[120] At the end of each tournament, a genetic algorithm, i.e., a mutable transition function, bred new teams from the successful teams to replace the less successful ones. Typically, in genetic algorithms, the performance of individual agents is measured and the fitness improved by cross-fertilisation and selection. In this case, the fitness of the team as a single system was measured by its aggregate score, and not that of individual agents. The simulation showed that although there was no communication between the players and no mechanism to attribute credit to individual players, the team performance improved as the individual behaviours were "selected" to complement each other, each unique agent playing its part in the team's strategy.

Each agent's transition function consisted of a set of rules represented by a string of characters, each taking the value of zero or one. The transition function took into account the agent's internal state, its nearness to the "ball", its geometric position, the position of other players in its immediate vicinity, and the position of the best-placed player in its team. At the end of each tournament, the transition functions (represented as strings of bits) of the most successful agents were bred to create new agents with new transition functions. How genetic algorithms breed is not strictly relevant to the theme of this book, but if you need an explanation look up "Genetic algorithm" in Wikipedia, the online encyclopaedia.

This simulation demonstrated how an agent's environment, i.e., the agent's position relative to the other "players" and the "ball", may be communicated to the agent and influence its behaviour. In addition to these skills and programming insights, the Team program clearly demonstrated that the team's emergent behaviour was derived from the interaction between the diverse behaviour of multiple agents.[120]

The remainder of this chapter describes a different agent-based, simulation program I wrote to demonstrate that such a program could simulate a co-evolutionary system. I called it SeeChange and it supports the attributes of change in socio-technical systems and the requirements of a simulation.

SeeChange allows the user to create a system consisting of many interacting components and to run the system over multiple time-steps in order to examine the behaviour of the system. The user creates agents, called sub-systems in See Change, with as many dimensions as needed. Each dimension has a name, type and value (or state). By default, the dimension types include integer, decimal and boolean, but the user can create new types with a name and a list of values, e.g., colour with the values red, green and blue, or height with the values tall, medium and short. The user can also create rules and functions to govern the component's behaviour and can save and load individual and groups of components, lists of type values and rules.

The connections between the components have a uni-directional influence weighting, an influence type, and a time-step increment. The influence weighting, in other words the degree of inter-component coupling, moderates the effect the influencing component has on the target component. The influence type permits the influence weighting to increase, remain constant or decrease over time, and the time-step increment governs how often the influence weighting is used. This last facility permits a variation in when the influence is applied; it can be at every time-step, every third time-step or, perhaps, every tenth time-step, reflecting the differing time that components take to change. The detailed operation of SeeChange is again not strictly relevant as it is the results that are most interesting.

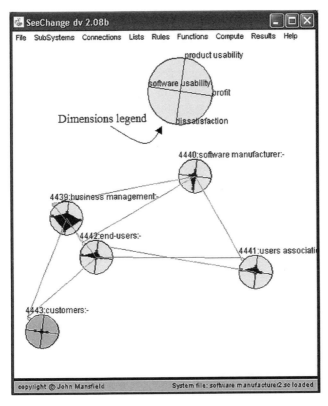

Figure 57. SeeChange — a system showing the influence the market players have on each other.

When the program is run in "Compute" mode, each of the components watches for "advertisements" from those other components that have an influence on it. The threshold and transition functions described in previous chapters are implemented by each component's associated rules and functions. It uses these rules and functions to determine if it should change and how it should change. A change in one component may influence or place a pressure for change onto other components; this may or may not cause a change in their dimensions. As a simple example of how SeeChange works, let us look at a system consisting of five components: a software manufacturer, the management of a business that uses the software to produce a product, a user of the software within the business, a customer of the business who purchases the product, and the software user's association. Each component operates in a design space of four dimensions: profit, dissatisfaction, software usability and product usability. Figure 57 shows the main screen of the simulation picturing the five components as grey circles; the large circle at the top is there to indicate the dimensions of the nodes. The lines of influence run from the centre of the influenced component to the edge of the influencing component.

The state of the system after each time-step computation can be displayed graphically (Figure 58) and in textual form (not shown).

The state/time diagram makes it considerably easier to see the variation in a dimension as the effect of a change propagates through the system. The SeeChange graphical results diagram extends the concept of a one-dimensional cellular automata state/time diagram to a multi-dimensional state/time diagram. The system dimensions, rules, functions and component attributes are displayed in separate windows, as in Figure 59.

Let me illustrate how SeeChange can be used to demonstrate the effects of the propagation of change through simple systems, before progressing to more complex systems.

The SeeChange Program and the System Structure

First, a simple SeeChange simulation is shown below, indicating how network connections influence change. In Figure 60 the top left-hand window shows 14 components, each with a single attribute, called "state", which can take two values — true or false.

The lines between the components indicate the connections that have a non-zero influence weighting. Each component is influenced only by the next component in a clockwise fashion. In this example, the protocol for change is very simple. The rules governing change are:

Rule 1: IF ANY state = true, THEN THIS state $\Rightarrow$ true
Rule 2: IF ANY state = false, THEN THIS state $\Rightarrow$ false

The IF ANY statement is interpreted as "**if any one, or more, of the influencing components**" meets the stated condition.

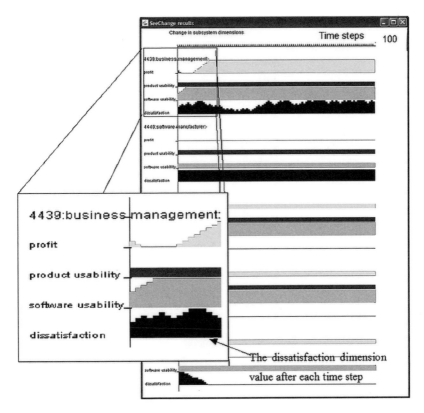

Figure 58. State/time diagram showing the change over time of each of the dimensions.

These rules are shown in the bottom right-hand window of Figure 60. Before the simulation started all the components were in the same state. One change was made by reversing the value of the component shown circled in bold in Figure 60, and the simulation started. The simulation was stopped after 50 iterations and the state/time results were observed in the upper right-hand window. As expected this simple system showed the changed state travelling endlessly round the system on a cyclic attractor.

The system was then changed from a regular network structure by making an influence connection across the system in small-world style. Thus, one component becomes influenced by two other components and the change propagation has two paths.

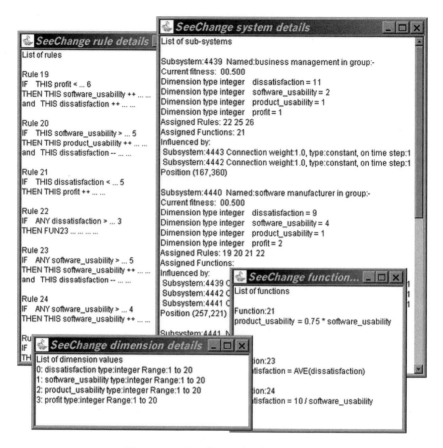

Figure 59. SeeChange details windows.

By observing the results window in Figure 61, it is clear that by creating a second path for the propagation of change, the behaviour of the system alters, the propagation of change again reaching equilibrium on a cyclic attractor. The third example in Figure 62 adds another connection, providing three paths for the propagation of change, travelling counter-clockwise, between node 1650 and node 2008. These are the indirect paths through all nine intervening nodes (i.e., taking 10 time-steps), the shortened path via node 1673 and two intervening nodes (i.e., taking 4 time-steps), and the direct path taking one time-step. Thus, the change arrives three times at node 2008, interfering with the previous pattern of propagation, altering the behaviour of the whole system, and causing the effect of the change to die away entirely.

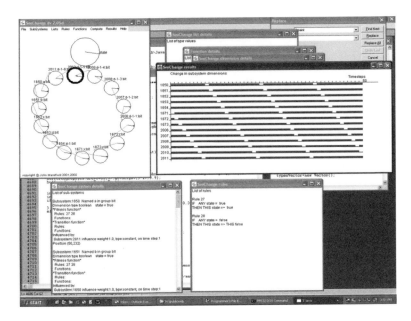

Figure 60. SeeChange screenshot 1.

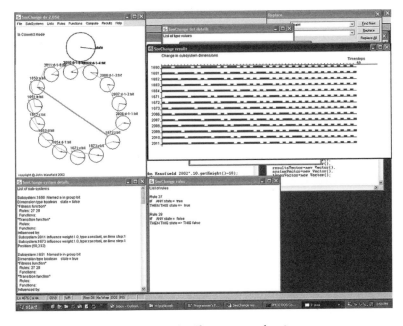

Figure 61. SeeChange screenshot 2.

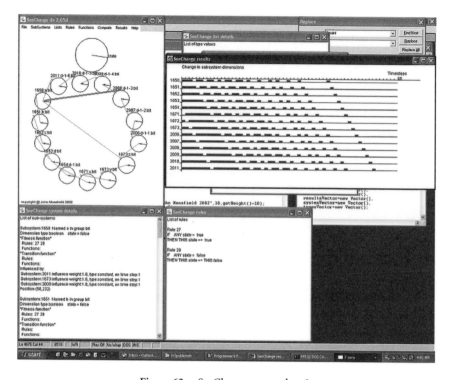

Figure 62.   SeeChange screenshot 3.

We can clearly see from these simple examples how the influence network structure of a system alters the propagation of change within the different structures.

A "Cellular Automata" Application of SeeChange

We can now check whether the propagation data derived from the one-dimensional cellular automata models transfers to this simulation. To do this, we extend the above application to a system where there are two influences on each component (Figure 63) and SeeChange is applied to emulate the cellular automata.

The cellular automata simulation indicated that the introduction of a second change while an initial change is still propagating had a high probability of creating deterministic chaos. This next simulation will test that result.

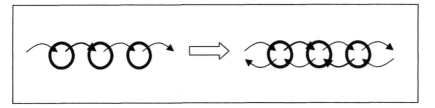

Figure 63. Extending the number of influences.

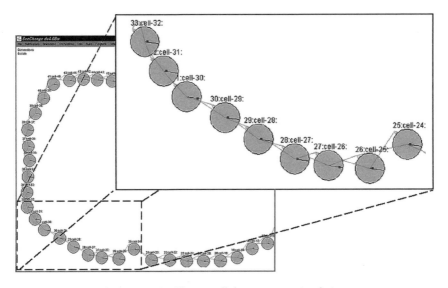

Figure 64. SeeChange cellular automata simulation.

A model system was created consisting of 70 components and, as above, each component has a single state that has two values. The 70 components are again arranged in the equivalent of a circle, each component having an influence connection to the neighbouring two components as in Figure 64.

Thus, at each time step, each component is affected only by two components. A screen shot of the simulation is shown in Figure 65.

The rule set for change is embodied in eight rules covering all possible combinations of the state of the central component and its neighbours. These rules are similar to the rules in the cellular automata known as WN 151. Actually, there are only six rules because two pairs of rules are

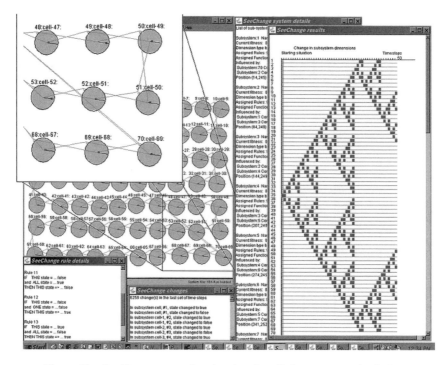

Figure 65. SeeChange screenshot showing the cellular automata simulation.

symmetric. A complete list of the rules and a brief explanation of their meaning are given below:

Rule 1	IF ALL state = true	THEN THIS state ⇒ false
	AND THIS state = true	

*i.e., if the state of **all** of the influencing components is true and the state of **this** component is true then make this state false.*

Rule 2	IF THIS state = true	THEN THIS state ⇒ false
	AND ONE state = false	

*i.e., if the state of **this** component is true and the state of **one** of the influencing components is false then make this state false.*

Rule 3	IF THIS state = false	THEN THIS state ⇒ false
	AND ALL state = true	

Rule 4	IF THIS state = false AND ONE state = false	THEN THIS state $\Rightarrow$ true
Rule 5	IF THIS state = true AND ALL state = false	THEN THIS state $\Rightarrow$ true
Rule 6	IF THIS state = false AND ALL state = false	THEN THIS state $\Rightarrow$ false

The system was initialised with every component set to false. When a single component was changed to true, the system produced a regular pattern identical to that of the WN 151 cellular automata.

A series of 20 simulations were undertaken to determine the effect of introducing a change after the system had run for five time-steps and the initial change had started to propagate. For each simulation, the system was first reset to a state where only the central component was in the true state, this component was designated "n". Initially the system was run for 105 time-steps (S1).

Subsequently, i.e., Simulation S2 onwards, the initial system was run for five time-steps and then one other component (in column $n + m$) was changed before running for a further 100 time-steps. As with the cellular automata simulations, these simulations produced behaviour that in some cases was regular but more often was deterministically chaotic. Statistically, the simulations are not significant but it was clear that four were regular and 16 were chaotic, and the set of simulations supported the conclusion drawn from the cellular automata simulation.

So it is clear that the introduction of another change into a system that is not in equilibrium propagates regularly until the second set of propagating changes impinge on the initial set of propagating changes. After which, the two sets of changes interfere and deterministic chaos can occur. Thus, the introduction of a change to any complex system that is not in equilibrium is likely to cause deterministic but chaotic behaviour, resulting in a highly uncertain system behaviour.

Small-World Application of SeeChange

If these two simulations are combined to form a small-world system, the system departs from the cellular automata model and starts

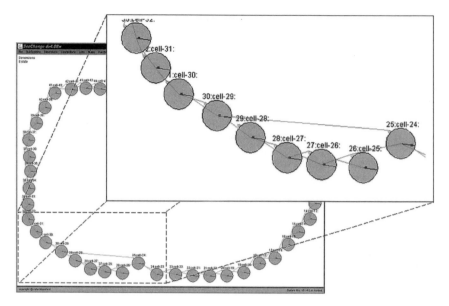

Figure 66.   Movement from a cellular automata model towards a small-world model.

to approach a real-world model. In Figure 65 the components were influenced only by their two nearest neighbours. In Figure 66 the link between component 25 and component 26 has been broken and a link has been made with component 30. Several of these new links were made and broken, and each time the SeeChange program was run for 3,000 time-steps.

As before, the results of these simulations can be shown as pictures of the changes from time-step to time-step (Figure 67) but this does not allow the results to be quantified in terms of cyclic or chaotic behaviour.

To provide this quantification the results have been recast in a different visualisation. As each row of component states contains only two states, black or white, a black state was represented by 1 and a white state was represented by 0. The row was then considered as a binary number unique to that particular system state, as in Figure 68.

These numbers were then converted to decimal and plotted as a graph to a logarithmic scale, as shown in Figure 69. After an initial settling period, the configuration system state in Figure 67 repeated, i.e., it fell into a cyclic attractor.

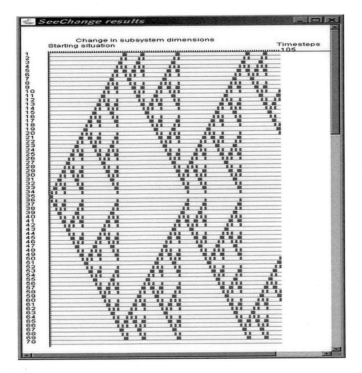

Figure 67. Time-step by time-step representation of the system state.

Figure 68. Binary representation of the system state.

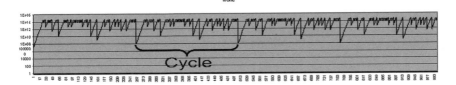

Figure 69. Graph of system states.

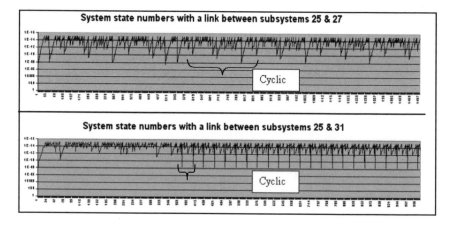

Figure 70.   Graphs of cyclic system states.

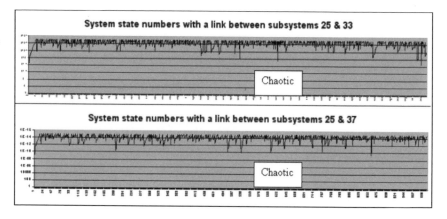

Figure 71.   Graphs of chaotic system states.

In the small-world configurations, where the link between components 25 and 26 was broken, the system with a link between components 25 and 27 was initially chaotic but eventually settled into a cyclic mode for the remaining 3,000 time-steps. The configuration of the system with a link between components 25 and 31 was again chaotic to start with, then settled into short cycles. Both are shown in Figure 70.

The remaining two sets of system states, i.e., the system with components 25 and 33 linked and the system with components 25 and

37 linked, remained in chaotic states throughout. It is clear from analysis of the data and inspection of the graphs that there are no cycles in Figure 71.

A set of similar simulations was then conducted but in this case, the link between 25 and 26 was maintained and an extra link was added. The simulations were run for 2,000 time-steps as it seemed clear that the interesting effects were in the earlier part of the time series and the later time steps simply repeated the pattern, whether it was cyclic or random.

Table 2 documents the length of the cyclic attractors in this set of simulations.

The change in a single link has a profound effect on the behaviour of the system because it alters the propagations paths from the initial change.

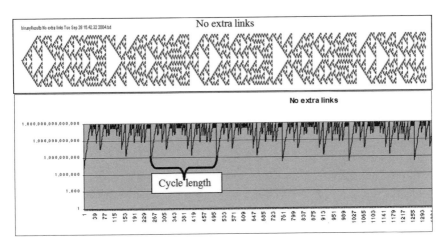

Figure 72.   Time-step representation and graph of system states.

Table 2.   Attributes of added link systems over 2,000 time-steps.

	No extra links	Extra link between 25 and 27	Extra link between 25 and 31	Extra link between 25 and 33	Extra link between 25 and 37	Extra link between 25 and 41
Cycle length	252	252	16	124	12	12

Two interesting inferences that may be drawn from these results are:

- The system behaviour is highly dependent on the propagation paths and hence on the structure of influence within a system;
- While a system may start with a deterministic chaotic behaviour, it can encounter a system state that pushes it into a cyclic behaviour (Figure 72); furthermore, the length of the cycle into which it falls can vary greatly.

Applying SeeChange to a Business Model

We shall now move to an even more realistic scenario where SeeChange is used to model the relationships between a software manufacturer, a software user association and three businesses. Each business is represented by a business management component, a customer component and an

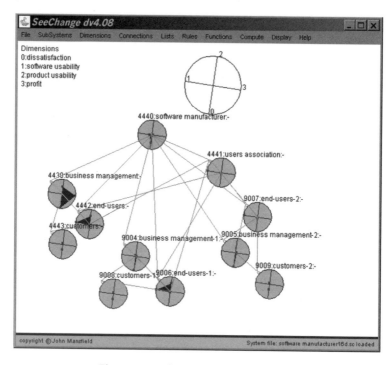

Figure 73.   A business system simulation.

end-user component. The end-user component represents the group of software users that develop the services provided by the business. The user association represents the interests of the end users of the manufacturer's software, nationwide. This system model is illustrated in Figure 73. Note that as the number of dimensions in the models will increase from here on, the key to the dimensions has changed from names on the circle to numbers. These numbers are associated with the dimension name at the top left-hand corner of the screenshot.

Note that in SeeChange the relationships are represented by lines of influence, a line going from the centre of the influenced component to the perimeter of the influencing component. The possible dimensions of the components are dissatisfaction, software usability, product usability and profit.

The simple rules in the system are:

Rule 37	IF THIS profit < 4	THEN ALL software usability ++
Rule 38	IF AVE dissatisfaction > 6	THEN THIS profit −−
Rule 39	IF AVE dissatisfaction > 8	THEN THIS profit −−
Rule 40	IF AVE dissatisfaction < 4	THEN THIS profit ++
Rule 41	IF ANY software usability < 3	THEN THIS dissatisfaction ++
Rule 42	IF THIS software usability < 5	THEN THIS dissatisfaction ++ AND THIS product usability −−
Rule 43	IF ANY product usability < 3	THEN THIS dissatisfaction ++
Rule 44	IF ANY product usability < 5	THEN THIS dissatisfaction ++
Rule 47	IF ANY software usability < 5	THEN ALL product usability −−
Rule 48	IF THIS software usability > 6	THEN THIS dissatisfaction −−

Rule 49	IF THIS product usability > 6	THEN THIS dissatisfaction −−
Rule 50	IF ANY software usability > 6	THEN ALL product usability ++ AND THIS dissatisfaction −−
Rule 53	IF ANY product usability > 6	THEN THIS dissatisfaction −−
Rule 54	IF ANY product usability > 8	THEN THIS dissatisfaction −−
Rule 55	IF AVE dissatisfaction < 2	THEN THIS profit ++
Rule 56	IF THIS product usability < 5	THEN THIS dissatisfaction ++
Rule 57	IF THIS product usability < 3	THEN THIS dissatisfaction ++
Rule 60	IF THIS software usability > 2	THEN ALL software usability ++

(Where ++ means increase the value by 1 and −− means reduce the value by 1.)

When this model is given a set of random values for the dimensions, the interactions can be played out over a number of time-steps. Figure 74 shows how the various dimension values wax and wane over 25 time-steps.

Taking the "end-user 2" component as an example, one can see that with an initially low software usability, dissatisfaction rises to the maximum possible by time-step 9.

This dissatisfaction is maintained until time-step 17, despite a rise in software usability. After time-step 17, dissatisfaction falls as software usability rises. In a real situation, the initial values could be estimated and useful conclusions drawn from the results. Where profit is waning for a particular component a change could be made, say to the level of software or product usability, and the result observed over a number of time-steps.

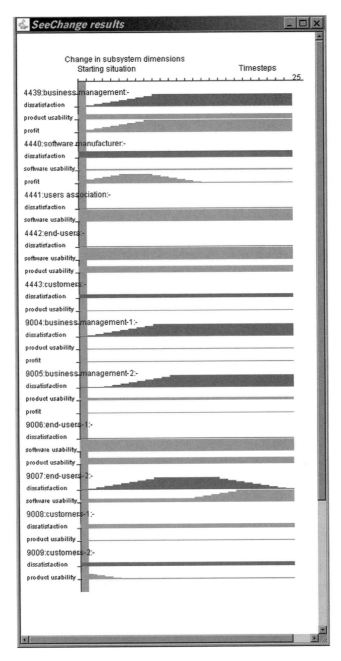

Figure 74. Changes in the business system dimensions over time.

Comparison of SeeChange with Commercial Tools

Social policy modelling is a typical complex, socio-technical problem that has generated many tools and has been attacked from many directions. SeeChange is a research tool but there are a number of commercial packages available with similar aims. However, many of these packages lead to mathematically-based "black box models". Even now, when many people use inscrutable systems without question, there is still a fear of the "black box model". Where modelling would be useful, the potential users often forego it because they mistrust the closed nature of the model. When it is used, the model builders must put considerable effort into making its structure and assumptions explicit. The models themselves are then unsuitable for direct use by decision-makers. There are software packages that aid decisions by representing uncertainty as probability distributions, calculating the propagation and displaying uncertainty results. But these too can be difficult for the decision-maker to use, as Millet Morgan and Max Henrion put it:

> "Spreadsheet applications let the user specify a correlation between two uncertain quantities as a way to express uncertain dependence. When there are more than two dependent variables, assessing the full correlation matrix is particularly difficult. Instead of using the abstract notion of correlation, it is often easier and more effective to model the underlying reasons for the dependence."[121]

Support for uncertainty is only one criterion in the selection of a modelling tool. Communication and ease of use for the decision-makers is another. Influence Diagrams were developed to represent the knowledge, uncertainties, objectives, and decisions of experts, decision-makers and stakeholders.[122] Some packages use an Influence Diagram that represents the system as nodes with arrows denoting the probability of one node influencing another. This form of the Influence Diagram is a knowledge representation tool to make more explicit the relationship between the various entities and assumptions in a decision-making scenario. The systems-dynamics concept introduced by Forrester[29] is an alternative concept. System-dynamic tools also represent the system using nodes and arrows, but to quote Morgan and Henrion again,

> "The two notations interpret the nodes and arrows differently. In systems dynamics, nodes represent stocks, sources, and sinks of conserved quantities, such as materials, water, money, or numbers of humans or other species. The arrows represent flows of these quantities, such as migrations, births, and deaths in populations. Other nodes represent valves, which are controls on flow rates. Influences, on the other hand, do not represent material flows — they represent knowledge and beliefs, about how the value of variables affects the value or probability distributions on other variables, which may reflect knowledge of material flows, or of other evidential relationships. For example, an influence diagram can express the relationship between a disease, its symptoms, and test results. Influence diagrams also identify decisions and objectives, explicitly.
>
> Second, systems-dynamics models are centrally concerned with cyclic relationships that represent positive and negative feedback loops. A conventional influence diagram is necessarily acyclic."[121]

None of these tools alone provides the facilities to predict change in complex systems, so SeeChange combines aspects of all these tools and adds further facilities not present in them. As noted above, SeeChange is a knowledge representation tool that graphically displays the system using nodes (components) and uni-directional connections. A SeeChange model shows the relationship between nodes, the effect of one node on another, and illustrates uncertainty but does not give it a quantitative measure. It is also a system-dynamic tool in that one of its most important aspects is its temporal facility; this facility allows the user to see the effect of changes over many time-steps. The most important point to note is that, unlike these tools, SeeChange nodes are multi-dimensional and hence their inter-node connection is multi-dimensional.

Additionally, many nodes combine to create the total environmental effect on a particular node. The resultant change is dependent on the intra-node, multi-dimensional coupling, itself requiring a complex trade-off and is, potentially, a multi-dimensional change. The multi-dimensional nature of a SeeChange node provides a mechanism to represent the components in the most natural way. The system is not a "black box model"; the dimensions of the components, the structure of the system and the relationship between nodes (both structural and rule-based) is explicit so decision-makers can satisfy themselves on the level of accuracy and assumptions of the model.

Insight and Understanding of System Behaviour

Simulations such as SeeChange can be used to gain an understanding of how particular systems behave when different forms of change affect them. SeeChange could be used in a way that parallels the way scientists use computer-based climate models to improve their understanding of the Earth's climate and to predict the impact of increasing concentrations of greenhouse gases. Scientists accept that the simulations are not entirely accurate or complete but they do give an insight into the complex subject of global warming. I shall discuss this popular topic in more detail in Chapter 11.

Insight into Interaction with Complex Systems

In Alaska, "statistically, someone dies from an aviation-related incident every nine days".[123] Operating complex systems such as aircraft in a complex terrain is an inherently dangerous affair. To reduce the risk, organisations are turning to computer-based simulations.

> "Officials at the Alaska Department of Veterans and Military Affairs have a new tool to take on the challenges that make the state the country's most dangerous place to fly. They are using $1.6 million worth of Space Imaging Inc.'s IKONOS satellite imagery to better navigate Alaska's 12 mountain passes. … They have contracted with E-Terra LLC to use the IKONOS imagery to develop flight simulation training modules, which include animations with 2- and 3-D viewing capabilities and cockpit control."[123]

And again, in the USA,

"Raytheon Co. has completed a prototype for a simulation and modeling system that eventually will help NASA and the Federal Aviation Administration evaluate revolutionary concepts aimed at improving traffic flow. A brainstorm-measurement machine, the NASA-commissioned tool will test ideas to determine their impact on the National Airspace System (NAS). Raytheon's goal is to provide a "plug-and-play" capability that enables users to plug in a scenario, watch it play out and evaluate its effects on the national airspace system."[124]

Such systems cost many millions of dollars but, in essence, they do a similar job as SeeChange in showing the propagating effects of making a change.

It is difficult to train operators to anticipate the result of making multiple operational decisions in factory process control, and this has resulted in a number of disastrous situations; Bhopal, Chernobyl and Three Mile Island are just three. Manufacturers are now using simulations to train their staff to handle hazardous situations without subjecting them to the danger of the real situation.

The military are another large user of simulations for planning and training. For example, in August 2004, it was reported in the press that the US Army had

"Two new simulators preparing 3rd Armored Cavalry Regiment and other soldiers for some of the most dangerous scenarios they could face when they return for their second deployment to Iraq. The Engagement Skills Trainer tests how soldiers respond to enemy fire from insurgents and the Convoy Skills Trainer gauges their responses to everything from vehicle breakdowns to direct fire while in a convoy formation."[125]

This chapter has given an overview of SeeChange and some of the insights it has provided in simulating change in complex systems. Note that SeeChange was written only as a research tool and does not have the resilience of a commercial tool. But in essence, it does permit the user to create a system consisting of many interacting components and to run the program over multiple time-steps in order to examine the system behaviour. It employs user-created rules and functions associated with each component to determine if the component should change and how it should change. Deliberate changes can be introduced concurrently and sequentially, and the system state examined time-step by time-step. The state of the system after each computed time-step can be displayed in textual or graphical form. The system can be run until it reaches an attractor (a single point or cyclic equilibrium) or deterministic chaos.

The simple simulations described in the chapter showed that system behaviour is highly dependent on the propagation paths and hence the structure of influence within a system. While a system may start with a regular behaviour, it can encounter a system state that pushes it into a chaotic or cyclic behaviour. Furthermore, the length of the cycle into which it falls can vary greatly. SeeChange models are open with all the assertions available for examination and amendment, providing more confidence in the model. Additionally, the rules and influence structure are flexible and can be added or modified as experience is gathered, mid-simulation if necessary.

9

What Do We Do When a Change is Indicated?

Messes, Problems and Puzzles

In the last chapter we saw how the co-evolutionary nature of change renders the forecasting of a system's emergent behaviour extremely difficult. In his own attempt to define the problem, Russell Ackoff suggests in *Redesigning the Future* that there are three types of system: a *mess*, a *problem* and a *puzzle*.[126] He defines a mess as a complex issue that does not have a well-defined form. With a mess, you do not even know what the difficulties are going to be. Typically, when an organisation proposes to create a large, complex information system, it starts with a mess. The mess can include technology, business, finance, government tax policy, contractors, customers or clients, and so on. For a successful outcome, all of these different system components must be identified and dealt with as a whole.

Next, Ackoff defines a problem as something that has a defined form or structure; i.e., it has components with attributes and variables, and we know something about how these variables interact. However, it does not have any single, clear-cut solution. As long it is a problem — in Ackoff's use of the term — it has many alternative solutions. The selection of the solution depends on the initial conditions, e.g., available technology, current business policy, available money, the party in government, who can supply what, to what schedule and what the customers or clients want.

Initially, much of this is unknown, so multiple solutions need to be considered.

Finally we have Ackoff's puzzle. A puzzle is a well-defined and well-structured problem with a specific solution. My experience indicates that when embarking on the creation of a large socio-technical system, many sponsors believe they have a problem when they actually have a mess, while engineers believe they have a puzzle when, if they are lucky, they actually have a problem. Pidd sums this up nicely using Ackoff's terminology:

> "One of the greatest mistakes that can be made when dealing with a mess is to carve off part of the mess, treat it as a problem and then solve it as a puzzle — ignoring its links with other aspects of the mess."[127]

Design by Increments

In my view, the way to approach system building, be it technological or social, is to design it by increments.

Design procedure

1. If it is clear that a change has to be made, then decide on the goal to be achieved. This may be increasing profit from a business process or reducing climate change. The goal is the target issue, not the change or the new system. An important part of the goal must be a quantitative or qualitative measure of success. Deciding on a general goal may be very difficult but without one it is pointless to do anything. Earlier I discussed a road safety example where the general goal of the traffic authority was to appease the road safety lobby by improving the safety on a particular major road. I will use this example again here. The measure of success or failure will be a reduction in the rates of road accidents.
2. Decide on what should not change. This is a step that is often ignored; perhaps later the inaction will be justified by stating that these assumptions were implicit. Do it explicitly, up front, and everyone will be clear that these assumptions are parameters of the change. In our

example, part of the goal is that the traffic flow and volume will not be reduced by more than a stated percentage. This percentage is one cost of the change.

3. At this stage we still have a mess, so identify all the stakeholders and other influences on the current system and the proposed general goal. These might be the road authority, the road safety lobby, the road users, the pedestrians, local businesses, the bus company, etc.

4. Move from a mess to a group of problems by breaking down the general goal into a number of possible courses of action, taking into account all the system influences and potential outcomes. These could be installing roundabouts or sets of traffic lights, building a new, wider road, starting a public education campaign, etc.

5. Now choose to solve one puzzle and make a specific change towards the general goal. This change should be significant and sufficient to have a measurable outcome. Let us say we will install a set of traffic lights at an intersection.

6. Allow the change to propagate until the environment either settles into equilibrium or settles into a cycle (oscillation) or is clearly chaotic (note: chaotic does not mean random). Observations might then show that the traffic lights improve the flow and safety of cross-traffic and pedestrians, but slow the traffic flow on the main road and result in congestions at peak times.

7. Evaluate the result of the change and decide if it has moved the environment towards the general goal. It is at this stage that difficult policy decisions need to be made.

8. If it is deemed a successful change, loop back and continue with the incremental design. If it is an unsuccessful change, in that it did not move the environment towards the general goal and/or it changed some entities or attributes that were not meant to change, then go back to step 4 and either roll back the initial change and try another of the options or leave the initial change and try to modify its outcomes. It may be that you need to go back to step 1 and explicitly modify some part of the general goal or the unchangeable attributes in order to move forward.

9. After each iteration, decide if the general goal has been satisfied. The degree of satisfaction determines whether or not to call things to a halt, i.e., it is your stopping rule.

Forecasting

There are three ways of forecasting the effect of all the time-steps between the current time and a specified time in the future:

(a) Wait until "it happens", i.e., experience all the effects of the changes in the system until the specified time becomes the current time. This has major disadvantages when there is a high cost if the system does not behave as anticipated.
(b) Paper and computer-aided analysis. This can take considerable time and can delay the decision to undertake the change beyond that allowed by operational constraints, and hence is often ignored in the decision-making process.
(c) Computer simulation. This is the fastest and most capable of aggregating the effects when dealing with many thousands of components. Its speed provides a "what if" facility to the decision-makers.

In options (b) and (c), possibly the greatest source of error is in the unknowable aspects of initial conditions; followed by the difficulty of determining the influences on a component and the difficulty of determining the trade-off and threshold within a component.

10

Implementing a System

Having decided on a strategy in the previous chapter, we now turn to the tactics of implementing a new system. I will use a computer-based information system because that is what I am most familiar with.

Early Architectures

Newer design techniques are becoming more common but for well over thirty years, computer-based systems have been designed based on the "Waterfall" model. While computer-based systems explicitly use this model, many other systems use it implicitly to create laws, regulations, medicines, company policy, etc. Winston Royce first described the concept in 1970 in a conference paper[128] although he did not name it, and as recently as 1995 the US Department of Defence was still saying (and may even still be saying):

> "DOD-STD-2167A/498, the current prevailing standard guiding software development, has been interpreted as mandating a specific process for use on all military acquisitions. This process is represented by the 'Waterfall' Model, which serves as the conceptual guideline for almost all Air Force and NASA software development."[129]

The "Waterfall" model stated, in essence, that requirements were to be written down and a design created. Appropriate hardware was then purchased, the software was purchased and/or written in-house and the system was integrated. This was, and still is, simplistic. In fact, there were always some, and sometimes many, changes made during implementation. In too many cases the implemented system was either not completed or, if it was designated complete, did not satisfy all the users' needs. You may recall I commented on this topic in Chapter 3.

More than a quarter of a century after Frederick Brooks[30] observed that change is a constant in socio-technical computer-based systems, there is still a reluctance to acknowledge that change will occur. Brooks noted that building computer-based socio-technical systems was unlike any other engineering projects, in that there is a need to accept that human creation of any such system starts with imperfect knowledge.

I would even go so far as to say that calling the process "engineering" is a mistake. It is no more engineering than sculpture is engineering; emergence and creativity play more of a part than do physics or chemistry. Sponsors and potential users of a proposed system are often vague when describing their needs and unaware of the possible benefits and pitfalls of the technology. Furthermore, system creation takes time and during this period the various users' needs may change — the emerging technical system and the human system join to form a complex, interacting, co-evolutionary socio-technical system.

The prevalent but simplistic view, as depicted in Figure 75, is that a computer-based system is a closed system created in isolation and that feedback from the system's environment may be neglected. The boundary is set too near and the mindset is that the world outside the machine is neglectable. It is not.

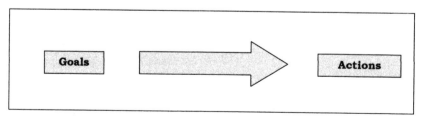

Figure 75.   Simplistic view of a computer-based system.

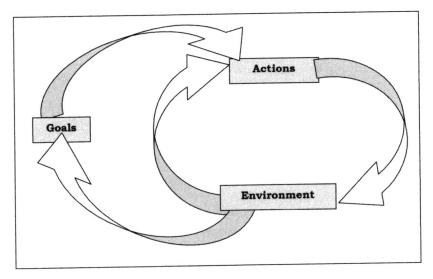

Figure 76. A more practical view of a computer-based system.

A more practical view expands the boundary and considers the immediate environment (Figure 76). This view accepts that installation of the system will alter, for example, the way people work.

Indeed, Manny Lehman is even more emphatic, saying,

> "Such installation changes the operational domain. The domain with a system installed and operational is different to that before installation. Installation also changes the application. Apart from the additional activity arising from operation of the system, there will be changes in the activities associated with the application. **If that were not so, there would be no point in installing the system.** [Author's emphasis in bold.] Thus the application and its intended operational domain that provided the initial inputs to the development process are changed by the output of that process."[130]

This view still neglects the most important aspect of the environment — the entities external to the immediate feedback loop. As Figure 77 illustrates, it is from these entities that the unexpected and unplanned feedback comes.

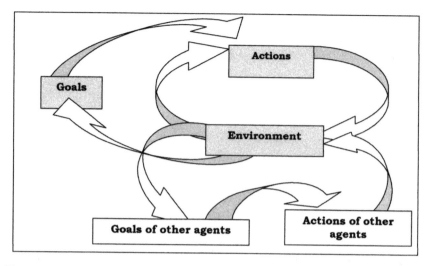

Figure 77. A more complete view of a computer-based system. (Diagrams are adapted from ideas in a paper by John Sterman.[78])

Because of this external feedback, any man-made system must always be imperfect and out of date when it becomes available for use, and this mismatch increases after it has been in use for a short while.

Mutability

What, therefore, are the factors that affect the mutability of a system? Stewart Brand has a view on architecture that helps to answer this question in the context of his type of socio-technical system, i.e., a building. In his book *How Buildings Learn*,[131] Brand observes that buildings change as people use them. He notes that the components of a building change at different speeds, for example, the structure changes slowly but the "stuff" (his word for furniture, etc.) changes quickly. This is mirrored in computer-based systems with multiple streams of events interacting with the system. Each of these external and internal event streams moves at its own pace and causes change to take place at the pace of the stream. User interfaces are exposed to a fast paced event stream, so user interfaces can change more rapidly than, say, the choice of operating system.

The operating system event stream is slow paced but change, when it comes, is major, akin to knocking out a wall in a building.

Furthermore, Brand makes a controversial contention that architects design to a fixed, superficial vision, concerned primarily with the building's appearance instead of its use. According to Brand, architects regard as perverse occupiers that have a changed vision when they come to occupy the building. Nevertheless, the buildings that users feel most comfortable in are those that can be easily adapted to their needs and can continue to change as time passes.

The Crucial Role of Time

Brand also states that "ALL BUILDINGS are predictions. All predictions are wrong".[131] What Brand is really saying here is that before you build a building, you plan for what is to happen and predict that the realised building will solve the problem it is being built to solve. However, because the information used in the original plan is imperfect and because buildings take time to build, the building when handed over to the users will, almost every time, fail to solve the problems that are now current. This is exactly paralleled in the building of large information systems.

Brand makes another key point that although buildings look mono-lithic and quite difficult to change, they are actually very plastic and malleable if designed appropriately. He argues that this is because a well-designed building is actually several co-existing and interrelated systems, designed so that changes to one system has minimal impact on changes to the others, i.e., they are loosely coupled. In a modern building the space-dividing walls are not structural, so they can be easily moved around; services are quarantined to particular parts of the building, so as not to impact on changes to the space plan; the structural framing of the building is separate to the skin, and so on.

All these changes allow the different aspects of the building — its look, it space plan, its services — to evolve independently. This in turn means that these different aspects can evolve at different speeds — perhaps the space plan every few years, and structure and skin in the order of decades or longer. In effect, Brand is saying that for a building to be adaptable, the coupling between layers should be minimised. This temporal flexibility makes good buildings malleable and adaptive to changing circumstances.

The relationship to computer-based and other socio-technical systems is obvious. Currently, many socio-technical systems seem to be designed the way architects designed buildings a hundred years ago — the space plan, structure, etc., were all tightly interwoven so that changing one required changes to all the others. There is a need to determine what the "layers" or "components" in system architectures should be, so that we can achieve the same flexibility, i.e., we should design for change.

Design of Computer-Based Systems

Some of today's architects of computer-based systems seek to reduce the inevitable imperfection in the original problem specification by gathering information as implementation proceeds, using a technique known as Spiral or Rapid Application Development. This technique creates a high-level prototype that is critiqued by the users, after which a more refined prototype is constructed and again critiqued by the users. The iterative process is continued until the users are satisfied that they have a useable system (or until the money runs out). What you then have is a system that fits the current environment and its problems as seen by its current users. This is a great improvement on the "Waterfall" method but it still does not take into account "environmental" changes over time and can be monolithic.

In the context of large, computer-based information systems, the computer system's environment is the user and the user's organisation. As the users become competent in using the information system, they often see new ways of doing things and dream up new things to do with the information. Additionally, the system attracts new users with different ideas of functionality. These new concepts change the organisation and its perception of what is required from the information system. So the computer system's environment changes; if these changes cannot be incorporated easily into the system, the users become frustrated and dissatisfied. To derive the expected benefits, the computer-based system and its users must continually co-evolve. Which raises the question that if "all predictions are wrong", how can we build systems that co-evolve with their users?

Design Strategy

Brand says that there is an escape from his pessimistic statement in that "Buildings can be designed and used so that it does not matter when they are wrong".[131] He suggests that the technique of scenario planning, long used by the military, can assist in building a building or any complex system that works well and continues to work well. Where a plan or design is based on forecasting what it will achieve, a strategy is required to deal with unforeseeable changes. With a good strategy, you always have manoeuvring room. Indeed, Brand's contention is "that many a building is a brilliant (or pedestrian) solution to the wrong design problem". We have already seen that this occurs in computer-based systems.

Brand also says, "Programming (in this case he means planning for a building) cannot accommodate perversity". This also applies to computer-based systems; however hard a computer-based system's architect or design team works to extract the needs of the client, the client invariably changes his mind once the design starts. The design team sees this as perverse but the client sees it as conforming to the real world, i.e., his or her own environment.

In using the scenario-planning technique to devise a computer-based system, I suggest the following process derived from Chapter 9 can be used; the steps may seem self-evident but my experience dictates that they are rarely all used together.

- Identify the major issue that makes it necessary to build a new computer-based system; this issue is the target not the new system.
- Explore the driving forces that will shape the whole environment that the system will run in.
- Identify a set of possible scenarios (including the scenario the system is expected to operate in) and work out the basic uncertainties. Think the unthinkable, identify events that might happen and that would have a significant effect, e.g., a terrorist bomb attack, a flood or a stock market crash.
- Revise the basic scenarios and flesh them out, name them.
- Work through a "day in the life" of each scenario.
- Cull them to no more than five, but leave in one or more of the more extreme scenarios.

- Devise a strategy that will satisfy the target and accommodate ALL the scenarios.
- In the light of the strategy, revisit the scenarios.
- Revisit the strategy.

Only now do you start looking at designing a system to satisfy the major issue. Participants in the strategy team should include people from a variety of backgrounds — business managers, clients or users, technical people, project managers, accountants, marketing, etc., but not more than 6–7 people including the team leader, plus a note-taker.

What shape should the design of large information systems take so that the technical system can co-evolve with the user's requirements? Well, hindsight is much more accurate than foresight, so the design must encourage "future hindsight", another Brand concept[131] meaning perpetual re-appraisal and adjustment. An information system's architecture should facilitate future hindsight.

For perpetual reappraisal and adjustment, you need an architecture that permits you to:

- Postpone some design decisions;
- Add and remove components;
- Delegate design power to the users and let them change things;
- Above all, acknowledge perpetual flux.

In short, you need an architecture that facilitates co-evolution. Microsoft's word processing application, Word, is a good example of an evolvable, component-based, complex system. It consists of a relatively small entry component, winword.exe, and hundreds of subsidiary components. Word's architecture allows it to be changed to suit a user's needs, whether they be novice or expert, using English or one of many foreign languages, and then to be changed again as the user's needs change.

11

Real-World Change: EUREKA Class

Let us now move away from hypothetical computer-based systems and look at an example of co-evolutionary change taken from the real world. The subject is real, but as I am going to discuss change within a class of warship, the description below has been fictionalised.

The EUREKA class consists of ten ships built over more than a decade. As time has passed, innovations and modifications have been incorporated into successive ships so that the class now consists of ten ships, all of which have a different configuration. Changing the ships to a similar configuration is a long and costly exercise.

Each ship is itself a very complex socio-technical system, so changes must be carefully considered in order to achieve the required result. Furthermore, each ship has an unending list of planned changes as defence policy and capability requirements change (Figure 78). Some of these changes have a small impact, say improving the accommodation in each of the seamen's messes, and some are large, like installing a new missile system. Some of these proposed changes are considered more urgent than others and they all take different lengths of time to implement. Changes that have been on the list for some time are well known and well planned; those that have been newly added to the list may have had little planning. Together, the list of changes and their interaction create a complex and dynamic network of events.

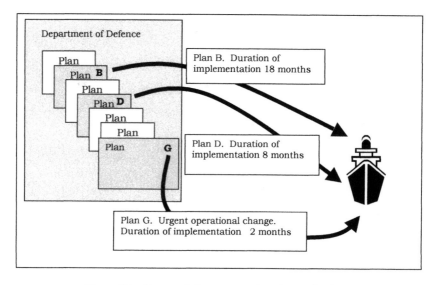

Figure 78.   Proposed changes at a given instant in time.

Aspects of the Change Process on a Ship

There are a number of aspects to be considered when a change is made to a ship. These include:

- Planned changes
- Changes currently underway — different changes take differing periods of time
- Priority of changes
- Unplanned changes rising out of planned changes, either because of unexpected emergent system behaviour or poor planning.

These changes affect:

- Current physical configuration
- Current properties — speed, radar signature, weight, size of crew, etc.
- Service life margins — weight and moment margins, mean time to failure, etc.

EUREKA Simple Simulation

First, we will create a model of the ship so that the change propagation process can be simulated in a less costly and less hazardous way. We start with a simplistic model of change in a EUREKA ship; a more realistic model is described later.

The model consists of three components in a design space of eight dimensions; the design space is illustrated in Figure 79 with arbitrary magnitudes. The problem addressed is to determine the effects of installing both a deck gun and a radar system on a ship.

In Figure 79, the small squares indicate the current value of the dimension within the range, denoted by the black line from the centre to the periphery.

The simple model in Figure 80 shows three components: planning, ship configuration and ship properties. In each component only the appropriate dimensions have a magnitude and the other dimensions are set to null. The ranges of safe ship properties are not part of the diagram but are included here for clarity; they are actually included in the model rules. The dashed arrows show that changes in the planning component influence the ship configuration, and changes in the ship configuration influence the ship properties. The model is now worked through as a paper "simulation".

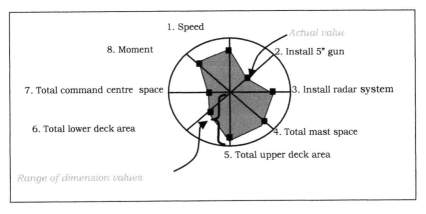

Figure 79.   EUREKA ship — simplified design space.

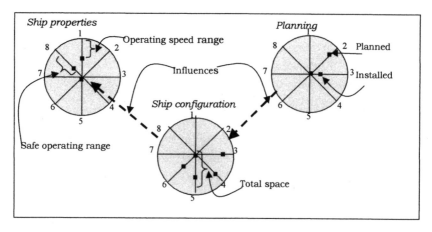

Figure 80.   EUREKA ship — simple model.

Independent Changes

First, the individual changes are considered independently and the results indicated.

In this paper "simulation", when the changes A and B (Figures 81 and 82 respectively) are considered independently, each result is as expected when the change was planned.

Now consider the changes A and B simultaneously.

The paper "simulation" in Figure 83 indicates that when the changes are considered together, they result in an unsafe situation and an inadequate top speed. An altogether unsatisfactory situation caused by the interference of the two changes.

Impact Analysis

In order to analyse the impact of a change, we need to note all the components and the influence they have on each other before transferring them to the simulation. For the purpose of this investigation, we divided each analysis into five parts to mirror the normal process:

- Planning is the initial part where the decision is made to add or modify a capability. (In Department of Defence terms, a capability is a high-level

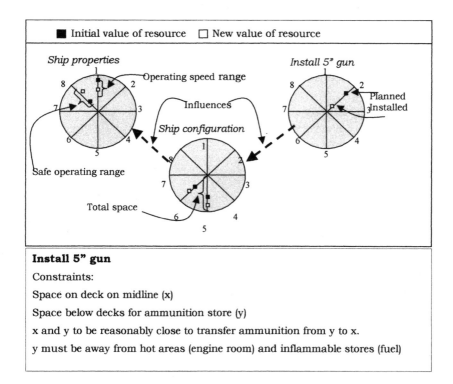

Figure 81.   Independent change A.

operational concept, such as support for an amphibious landing or detection of hostile aircraft at a great distance.) This part also includes the defining of the Concept of Operations (i.e., how the personnel, hardware and software will be used to achieve the aim of the capability) and planning the procurement of the resources.

- Procurement is the process by which the trained personnel, hardware, and software are obtained and installed.
- Attributes detail the features of the personnel, hardware, and software that will be introduced onto the ship.
- Derivatives, in this context, are the requirements that will influence the current configuration of the ship; typically, these are electrical power consumption, area and/or volume, weight, and placement, but there are many others.

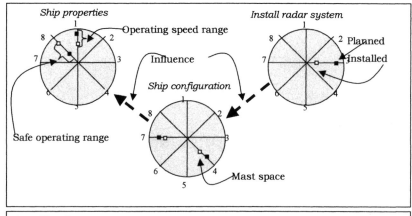

Figure 82.   Independent change B.

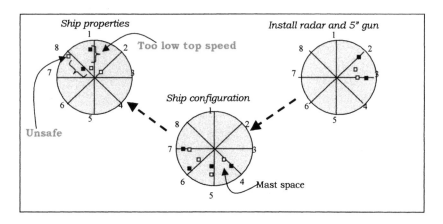

Figure 83.   Concurrent change.

- Operating constraints describe a number of vital properties the ship requires to effectively carry out all of its desired capabilities. For example, stability (centre of gravity and turning moment constraints) is necessary for the vessel to operate on high seas without capsizing.

There is often more than one capability change in the pipeline and each of these change processes interact on the actual vessel, hence with each other. To encompass the complexity of all this activity in a model requires a knowledge management strategy similar to that indicated in Figure 84, which shows a schematic of the knowledge management of only one change.

A manual impact analysis requires a chart (Figure 85) to guide the analysis. I developed this chart to indicate how the decision to add or

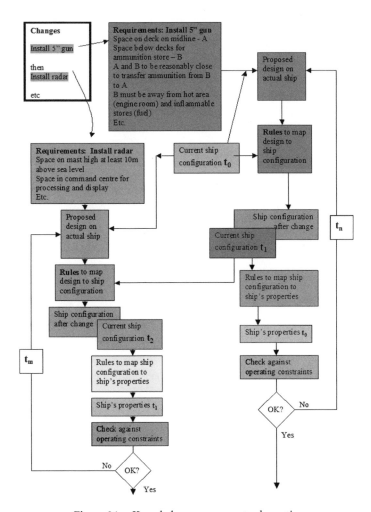

Figure 84.   Knowledge management schematic.

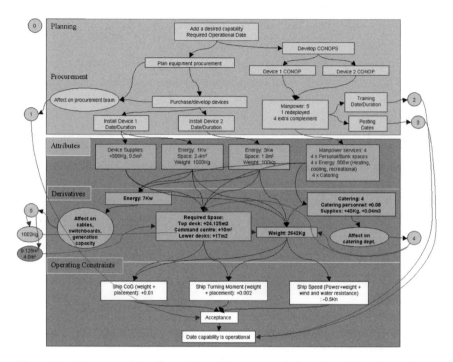

Figure 85. Impact analysis chart (the numbers around the edge of the chart refer to linked charts that show more detail).

modify a capability flows down into a new configuration for the ship, and indicates whether the changed operating constraints are acceptable.

Because of the dynamic nature of the ship's configuration, it is still not possible, even with an impact analysis and a knowledge management strategy, to fully appreciate the effect of the introduction of a particular capability. It is therefore necessary to utilise a simulation program to see the propagation effects of changes. Ideally, the program will permit "what if"-style experiments to determine if there is a propagated effect that breaches the operating constraints, and if the order of introducing changes makes a difference.

A textual analysis of the changes and a model of changes on a single EUREKA ship are shown on the next page.

A Model of Change on a EUREKA Ship

Before proceeding to the simulation, let us return to the protocol for change and describe some aspects of the protocol in terms of an impact analysis of changes on the EUREKA ship. The protocol discussed earlier may be summarised this way: a change will occur if, and only if, the pressure for change is greater than the resistance to change. The pressure for change is the aggregate of the goal of the component and the sum of the influences from the other components. The resistance to change is the aggregate of the risk engendered by a change to this component, the cost of a change to this component and the system level control that determines whether change will take place or not.

The emergent behaviour of the EUREKA ship, as a complex system, is determined by the cascade of changes engendered by one or more initially independent changes as the effects are propagated through the many components of the ship. The list of planned changes creates a pressure for change and, in the simplest case, cost may simply be represented as consumption of resources — space, electrical capacity, etc., and change control may be represented by the constraints of maintaining a balanced and properly operating ship. The desired state of the ship is defined by a set of operating requirements and resources, hereafter called the ship's operating constraints. These can include deck space, top speed, seakeeping, relative position of the centre of gravity, turning moment along the major axis of the ship, etc.

If after applying multiple changes the propagation of effects causes the operating constraints to be exceeded, then one change may need to be rolled back to permit other more urgent changes to be implemented, affecting the project budget and schedule. The implementation and subsequent rollback is a costly waste of expenditure. The exceeding of operational constraints can also generate another "unplanned" change, for instance, the installation of a larger electrical capacity. This new change may have a higher priority than existing planned changes and hence will inevitably change the installation schedule; it may consume resources that would have been available for other planned changes. This may generate other "unplanned" changes, and they may propagate to encompassing systems.

Take, as a fictitious example, the installation of a 5-inch gun on the foredeck. The ship was designed to have a gun on the foredeck so space was left for it on the deck, but the size of the gun has been increased to provide the capability of maritime artillery support to troops on shore. The gun is to be installed during a two-month period when the ship will be in dock before leaving for a distant war zone. This period is short, so a tight schedule is required. As much as possible of the work is to be carried out before the ship docks, and the equipment is to be ready for installation on the dockside when the ship arrives. Thus, in terms of the change protocol, the goals of the gun component are to improve the ground troops support capability, and that the capability should be operational before the ship's scheduled date of departure.

The influences come from the stakeholders:

- the Defence Force Chief and the Army Chief who want the capability in place,
- the Navy Chief who wants the ship back at sea on schedule,
- the Defence material organisation, who have to tender for the gun, the ammunition and the installation, accept tenders for the material and installation work, write contracts and control the budget,
- the armament factory that is to provide the gun,
- the ordnance factory that is to provide the ammunition,
- the Defence Training Branch who wants several members of the crew off the ship to be trained in the gun's operation before the installation takes place,
- the crew members who are currently using the designated ammunition store as a recreation space.

The risks are:

- due to delays, another ship is still occupying the designated dock when the ship arrives, so the material, cranes and other equipment have to be moved to a less suitable dock,
- the preparatory work is not completed in time,
- the material is not delivered on the dock on time,
- the installation does not go smoothly, and either the installation is incomplete when the ship leaves or the ship's departure is delayed,

- there will be labour problems on the dock,
- there are no personnel to be trained or there will be training problems so there will be no trained gun personnel,
- interference from the effects of other concurrent changes on the ship will cause the schedule or budget to overrun, and
- the changes due to the installation of the gun will interfere with other concurrent changes.

The costs are:

- the loss of the military advantage if the capability is not available on schedule,
- the contractual costs and any overruns,
- the stress on the crew, suppliers and dock staff to get the capability available on schedule.

The control is exercised by:

- The Department of Defence (including the Defence material organisation), the ship's captain, the shipyard management, and labour.

Applying the protocol qualitatively, the stakeholder influence from the stakeholders is strong and since the aim of the project is aligned with this influence, the pressure for change is strong. The risks are understood and plans have been made to minimise them, the costs are understood and accepted, and the control accepts the need for the change so the resistance is small. After weighing the goals, influence, costs and risks, etc., the decision is made to make the change.

The transition period starts when the decision has been made and lasts until the capability is operational. The propagation effects start with the extra weight of the larger gun requiring further structural support under the foredeck. This requires pipes, electrical, computer and sensor cables be moved before the support can be put in place. Rerouting the sensor cables increases their length so repeater amplifiers need to be added. Once the gun is in place, extra electrical cables are required to power its electrical components, and the cables are run to the nearest switch panel, the forward switch panel.

As the gun is of a larger calibre than originally designed, the lift that takes the ammunition from the store to the foredeck needs to be enlarged. Because of the larger ammunition, the enlargement of the lift and extra support for the gun, the amount of ammunition that can be stored is reduced by 15%. Thus, the planned capability is reduced.

The crew that formerly used the ammunition store for a recreation space are forced to join their mates in the messroom, causing over-crowding. The seamen designated to be trained on the gun could not be made available until the ship docked, so they are denied the training opportunity and a trained gunnery seaman has to be added to the ship's company. These events affect the morale of the crew but they are mollified by the promise of air-conditioning in all the messes.

Concurrent with the decision to install the gun, another part of the Defence material organisation proposes to take advantage of the ship's presence dockside to make a long-awaited upgrade to the long-range air search radar system. This version of the radar system is heavier and more powerful than the one it is to replace on the masthead.

The goals of the new radar system are to improve the range and reliability of the ship's long-range air search radar capability, and that the capability should be operational before the ship's scheduled date of departure.

The influences come from the stakeholders:

- the Navy Chief who wants the long-range search capability but also wants the ship back at sea on schedule,
- the Defence material organisation who have already written contracts for the procurement and installation and do not want the budget to be exceeded,
- the factory that is to provide the radar system,
- the ship's captain who is pressing for trained radar operators on the new equipment.

The risks are:

- escalation in the cost of the contracts as they were awarded some time ago and have been in abeyance awaiting a suitable time for the installation,

- the preparatory work is not completed in time,
- the material is not delivered on the dock on time,
- the installation does not go smoothly and either the installation is incomplete when the ship leaves or the ship's departure is delayed,
- posting or training will be delayed so there will be no trained radar personnel when the radar system becomes operational,
- interference from the effects of other concurrent changes on the ship will cause the radar system's schedule or budget to overrun,
- the changes due to the installation of the radar system will interfere with other concurrent changes.

The costs are:

- the continued danger to the ship if the new radar system's improved range and reliability are not available on schedule,
- the contractual costs and any overruns,
- the stress on the crew, suppliers and dock staff to get the capability available on schedule.

The control is exercised by:

- the Defence Department, the ship's captain, the equipment suppliers, the shipyard management, and labour.

Applying the protocol qualitatively, the stakeholder influence is strong and the goal is aligned with the influence, so the pressure for change is strong. The risks are understood and plans have been made to minimise them, the costs are understood and accepted, and the control accepts the need for the change so the resistance is small. The decision to make the change was made some time ago. Again, the transition period starts when the decision has been made and lasts until the capability is operational.

The propagation effects start with the extra weight of the radar system, requiring the mast to be strengthened. The electrical power cables need to be upgraded to carry the higher power requirements of the more powerful radar. The previous cables ran from the forward switch panel and the extra electrical load was within the parameters of the panel when the installation was planned. However, the gun installation has taken the switch panel to the limit of its power-carrying capacity, so the switch panel and the cable supplying it need to be upgraded back to the generating equipment.

The extra power requirements of the gun, radar system, and air-conditioners take the electrical generating capacity over its limit. The solution is to either install a larger generator or an additional auxiliary generator.

The goals for the new electrical generating system are to improve the ship's electrical generation capacity, and that the equipment should be operational before the ship's scheduled date of departure.

The influences come from:

- the ship's Chief Engineer who wants the extended facility,
- the ship's Captain who wants the ship back at sea on schedule,
- the Defence material organisation who have no provision for the procurement and installation, and do not want the budget to be exceeded,
- the suppliers of electrical generating equipment.

The risks are:

- the Defence material organisation will be unable to fund the upgraded electrical capacity,
- there is insufficient space to install a larger or auxiliary generator,
- the suppliers will be unable to provide a generator in the required timescale,
- the inability to provide sufficient power to operate both the gun and the radar system simultaneously,
- the installation does not go smoothly and either the installation is incomplete when the ship leaves or the ship's departure is delayed,
- interference from the effects of other concurrent changes on the ship will cause the budget to overrun,
- the changes due to the installation of the generator will interfere with other concurrent changes.

The costs are:

- the additional contractual costs and any overruns,
- the stress on the crew, suppliers and dock staff to get the capability available on schedule.

The control is exercised by:

- The Defence material organisation, the ship's captain, the equipment suppliers, the shipyard management, and labour.

Applying the protocol qualitatively, most of the stakeholder influence is strong for installation and since the goal is aligned with this influence, the pressure for change is strong. But the Defence material organisation has not set aside funds for increased electrical capacity and insists that they will have to come from another budgeted project. In addition, space is tight on the ship and the designer has left no space on the lower deck for a larger or an extra generator. These influences resist the change. The risks are understood and plans have been made to minimise them by installing an auxiliary generator on the middle deck, the costs are understood and their magnitude agreed on, but the Defence material organisation are resisting the provision of funds. The control accepts the need for the change but as funding the change is difficult, the resistance is moderately strong. The decision is made to make the change. Again, the transition period starts as soon as the decision has been made, but in this case lasts beyond the time when capability is operational due to the redirection of funding. The Defence material organisation has withdrawn funding for an upgrade to the command and control system software to offset the funding of the auxiliary generator. This means the new radar system has to be operated in stand-alone mode and is not integrated with the command and control system.

Because of space problems and because the generator gets hot and needs ventilation, the fuel tank for the auxiliary generator has had to be separated from the generator and placed against the bulkhead of the ammunition store. It is connected to the generator by a flexible pipe.

The gun and the radar system are installed to schedule with only minor budget overruns due to their installation interaction. The auxiliary generator was not available until late in the installation schedule so sea trials of the gun were postponed until the ship was on its way to its next posting in the distant war zone. The shipbuilder's engineers informed the Captain during their final briefing that the new gun, ammunition, generator, radar system, and extra air-conditioners had raised the centre of gravity of the ship but not to a dangerous extent. Two trained radar oper-ators and a radar technician were added to the ship's company. These extra seamen and the gunnery seaman require extra accommodation and cater-ing supplies. As the catering store is already at its maximum capacity, this extra consumption will reduce the time the ship can stay at sea, restricting its range.

With all new equipment accepted and working to the supplier's spec-ification, and all training completed, the ship leaves the dock. A day or two out in calm open water the gun was trialed in calm seas. It was found that when the gun was fired across the long axis of the ship, coupled with the extra weight of the radar system at the top of the mast and the rise in the centre of gravity, a dangerous turning moment was generated. Thus, if the gun is used broadside in anything but calm seas there is a possibility of capsizing.

The effect of the two acceptable changes has propagated throughout the ship and led to a potentially unsafe situation; however, the situation can be avoided by not firing the gun broadside, an unfortunate but not catastrophic situation. Additionally, because of the adverse Defence Enquiry Finding into a recent major shipboard fire, the Navy Chief has now promulgated a regulation that no fixed equipment on a ship should have a flexible fuel line. The fuel line to the auxiliary generator cannot be replaced at sea; it would be too dangerous to cut and weld piping with ammunition in the store on the other side of the bulkhead. If the fuel tank is drained to comply with the new regulation, the generator will not run and neither the gun nor the radar system can be used. This is now an unsustainable situation. The Captain has no choice but to return to port and the ship currently in the war zone is told to remain on station until her replacement is ready, thus starting another chain of propag-ating changes.

The scenario described above is plausible but, of course, hypothetical. Nevertheless, it could happen, and something very similar did happen to the Swedish ship *Vasa*, which capsized on its maiden voyage in 1648.[132] Further, this scenario is a far simpler set of inter-related components than occurs in practice. We can infer that this is a real example of the uncer-tainty of forecasting the effects of changes in a complex, socio-technical system.

A SeeChange Model of Changes to a Ship

The major problem with the previous descriptive modelling is obvious; writing is serial so it is not possible to deal with all the potential interactions in a stream of narrative. A better, quantitative method is to model the physical ship, the interaction of its components and the emergent behaviour following changes. A computer-based model permits all the interactions to be taken into account as they happen and the emergent behaviour to be viewed over time. SeeChange was designed and crafted to support this type of simulation. The rest of this chapter describes a simulation of the problem described above using SeeChange and the emergent behaviour encountered during the simulation.

Physical Model

Basic ship flotation mechanics

In order to understand the hypothetical EUREKA ship problems, I need to digress into a very brief introduction to ship flotation mechanics. This information is from the US Navy website: http://web.nps.navy.mil/~me/tsse/NavArchWeb/1/module2/basics.htm.

Certain points in the ship are described as centres and they are fundamental to ship behaviour at sea. The most important of these are the centre of gravity, the centre of buoyancy and the metacentre. I will define these centres below and show how they interact.

Centre of Gravity (G). Though the weight of the ship is distributed throughout the ship, it can be considered to act through a single point called the centre of gravity. If the ship were to be suspended from a single thread, that thread would be connected at the centre of gravity for the ship to remain upright and on an even keel. The weight always acts vertically downward through the centre of gravity. The location of the centre of gravity of a ship is solely a function of weight distribution within the ship. It is in a fixed position for each condition of loading of the ship, but moves whenever there is a weight addition, removal or movement within the ship.

Centre of Buoyancy (B). The centre of buoyancy is the geometric centre of the submerged hull. The force of buoyancy acts vertically upward through this centre. When the ship is at rest, with or without a list, the centre of buoyancy is usually directly below the centre of gravity. As the ship is disturbed, the centre of buoyancy moves to the new centre of the submerged hull. The force of buoyancy then acts vertically upward through the new centre of buoyancy. When the centres of gravity and buoyancy are not aligned vertically, the forces of gravity and buoyancy acting through their respective centres tend to rotate the ship.

Metacentre (M). The metacentre is an imaginary point that is of prime importance in stability.

When the ship is inclined to small angles, the centre of buoyancy moves away from the centreline. The metacentre is the intersection of the line of buoyant force acting vertically through the new centre of buoyancy and the inclined centreline of the ship. In a stable ship, the metacentre lies above the centre of gravity.[133]

From Figure 86, you can see that when a ship is in equilibrium the centre of buoyancy is directly below the centre of gravity. When a ship keels over because of wind, waves or other forces, a righting force is then caused by gravity pulling down on the hull, effectively acting on its centre of gravity,

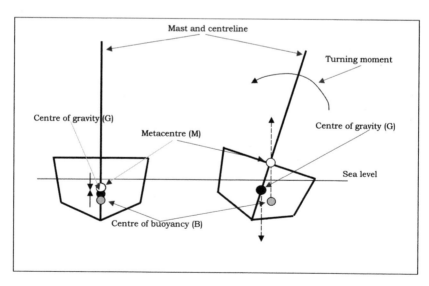

Figure 86.   Relationship of centre of gravity and centre of buoyancy.

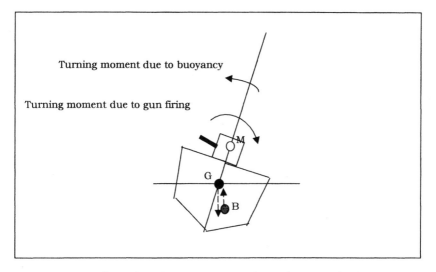

Figure 87. Relationship of turning moments due to the gun and to buoyancy.

and the buoyancy pushing the hull upwards, effectively acting on the metacentre. This creates a turning moment that rotates the hull upright again and is proportional to the horizontal distance between the centre of gravity and the metacentre (see Figure 87). The metacentric height is important because the righting force is proportional to the metacentric height multiplied by the sine of the angle of heel. A large metacentric height gives a ship a short roll period and good stability; on the other hand, a ship with a small centre of gravity to metacentre distance (G to M) will have a long roll period, and if the G to M distance is too small the ship will be at risk of capsizing in rough weather.

If a ship has a high centre of gravity, its G to M distance is small and it does not recover well from waves and wind that cause it to keel over. This tendency is exacerbated if there is excessive weight high up on the mast, such as a heavy radar system at the top of the mast. If another event, such as firing a gun, adds a further turning moment to the long roll period there is a possibility that the ship will capsize.

Taking these facts into consideration, a model of a ship was devised with weights and distribution of equipment in approximate proportion to that of a ship (Figure 88).

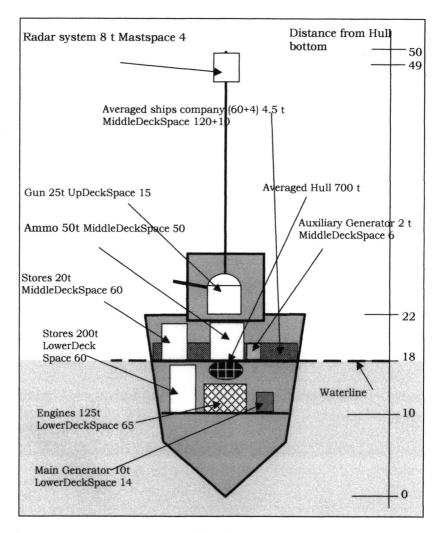

Figure 88.   A simplified physical model of a ship.

The data from this physical model was then used to create a SeeChange model. In this model, the changes all propagate towards the same end-point so the order of events is unimportant. However, to clarify the results, the changes were noted when the radar system, gun, ammunition and extra seamen was introduced sequentially.

I have not included here the details but all the attributes of the ship and the ship system components were given names and values, e.g., the engines were given the obvious name "Engines" and a weight of 125 tons, placed centrally 10 metres from the hull bottom, occupying 65 square metres of lower deck space (LowerDeckSpace). Remember all these attributes are fictitious.

Some of the large, user-defined rule set is shown below to give you an overview of the complexity of a required rule set for the real-world situation.

Rule 1	IF ANY Weight[t] > 0	THEN FUN1
Rule 2	IF ANY ElectricalPower [kW] > 0	THEN FUN2
Rule 3	IF ANY Speed[Kn] < 27	THEN THIS InSpecifiedSpeedRange $\Rightarrow$ false
Rule 4	IF ONE TurningMoment > 60	THEN THIS SafeTurningMoment $\Rightarrow$ THIS false
Rule 5	IF ONE Speed[Kn] > 27	THEN THIS InSpecifiedSpeedRange $\Rightarrow$ true
Rule 7	IF ANY Weight[t] > 0	THEN COG (COG is a function that computes the centre of gravity of the target component and all the connected components)
Rule 8	Discarded.	
Rule 9	IF ANY UpperDeckSpace [m^2] > 0	THEN FUN3
Rule 10	IF ANY MiddleDeckSpace [m^2] > 0	THEN FUN4
Rule 11	IF ANY LowerDeckSpace [m^2] > 0	THEN FUN5

Rule 12	IF ANY MastSpace[m] > 0	THEN FUN6
Rule 13/ 14/15	Discarded	
Rule 16	IF ANY Provisions[t] > 0	THEN FUN9
Rule 18	IF ANY ElectricalConsumption [kW] > 0	THEN FUN11
Rule 19	IF THIS ElectricalConsumption [kW] > THIS ElectricalPower[kW]	THEN THIS SufficientElectricalPower $\Rightarrow$ false
Rule 20	IF THIS ElectricalConsumption [kW] < THIS ElectricalPower[kW]	THEN THIS SufficientElectricalPower $\Rightarrow$ true
Rule 22	IF ANY Speed[Kn] > 27	THEN THIS InSpecifiedSpeedRange $\Rightarrow$ true
Rule 23	IF ANY CentreOfGravity > 16	THEN THIS SafeCentreOfGravity $\Rightarrow$ false
Rule 24	IF ANY CentreOfGravity < 16	THEN THIS SafeCentreOfGravity $\Rightarrow$ true
Rule 25	IF ALL UpperDeckSpace $[m^2]$ < 240 AND ALL MiddleDeckSpace$[m^2]$ < 260 AND ALL LowerDeckSpace$[m^2]$ < 260	THEN THIS InSpaceRange $\Rightarrow$ true
Rule 27	IF THIS InSpaceRange = true	THEN THIS InSpaceRange $\Rightarrow$ false

Rule 29	IF ANY EnginePower [kW] > 0	THEN FUN12
Rule 30	IF ANY TurningMoment > 0	THEN FUN13
Rule 31	IF ALL TurningMoment < 60	THEN THIS SafeTurningMoment $\Rightarrow$ true

Functions

A number of built-in and user-defined functions supplemented the rules. The functions available to be assigned to a rule or component were:

FUN1 Weight[t] = SUM(Weight[t])

FUN2 ElectricalPower[kW] = SUM(ElectricalPower[kW])

FUN3 UpperDeckSpace[m^2] = SUM(UpperDeckSpace[m^2])

FUN4 MiddleDeckSpace[m^2] = SUM(MiddleDeckSpace[m^2])

FUN5 LowerDeckSpace[m^2] = SUM(LowerDeckSpace[m^2])

FUN6 MastSpace[m] = SUM(MastSpace[m])

FUN7 — Discarded

FUN8 Weight[t] = SUM(Weight[t]) + SUM(Provisions[t])

FUN9 Provisions[t] = SUM(Provisions[t])

FUN10 Speed[Kn] = THIS(EnginePower[kW]) / THIS(Weight[t])

FUN11 ElectricalConsumption[kW] = SUM(ElectricalConsumption[kW])

FUN12 EnginePower[kW] = SUM(EnginePower[kW])

FUN13 TurningMoment = SUM(TurningMoment) + SUM(MaxTurningMomentFired)

Structure of the SeeChange Model System

Initially, the system structure looked like the screenshot in Figure 89.

The system structure has four layers as shown in Figure 90. In each time-step, Layer 1 components (OperationConstraints) are influenced by

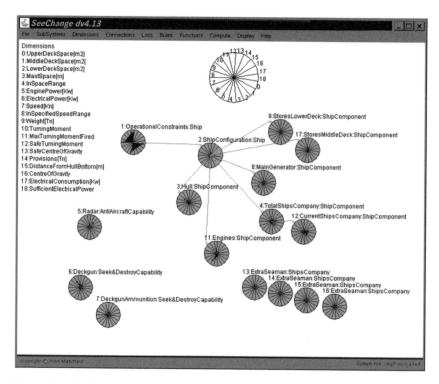

Figure 89. SeeChange model with the ship components and the initial influence connections.

Layer 2 components (ShipConfiguration). Simultaneously, Layer 2 components (ShipConfiguration) are influenced by Layer 3 components (Hull, TotalShipsCompany, MainGenerator, Engines, StoresLowerDeck and StoresMiddleDeck) and Layer 3 components (TotalShipsCompany) are influenced by Layer 4 components (CurrentShipsCompany).

To allow all the changes to propagate up the layers, the Compute time step has to be run three times. Then, for example, the influence of the weight and other dimensions of the CurrentShipsCompany propagate to TotalShipsCompany in the first time-step, to ShipConfiguration in the second time-step and to OperationalContraints in the third time-step. As there are no feedback influences in this simple system, it is now in equilibrium and further time-steps do not alter it.

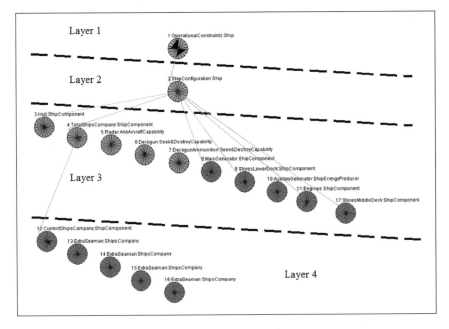

Figure 90. The four layers in the system.

Simulation results

The following subset of the simulation data will give you an idea of the sort of results achieved.

Initialisation

After three time-steps the following changes had been made.
ShipConfiguration Weight[t] changed to 1239.5
ShipConfiguration ElectricalPower[kW] changed to 120.0
ShipConfiguration DistanceFromHullBottom[m] changed to 15.838
ShipConfiguration MiddleDeckSpace[m^2] changed to 180.0
ShipConfiguration LowerDeckSpace[m^2] changed to 139.0
ShipConfiguration ElectricalConsumption[kW] changed to 90.0
ShipConfiguration EnginePower[kW] changed to 35720.0
ShipConfiguration CentreOfGravity changed to 15.838

ShipConfiguration TurningMoment changed to 50.0
ShipConfiguration Speed[Kn] changed to 0.0.

Installing the Radar system

The Radar system was then installed (i.e., linked to the ShipsConfiguration component by an influence connection) and three ExtraSeaman components were added (linked) to the TotalShipsCompany component. The system now looked like Figure 91.

After a further three Compute time-steps the following changes had been made.

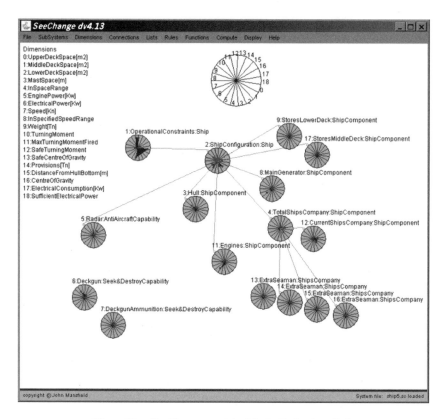

Figure 91.   SeeChange model with the Radar installed.

ShipConfiguration Weight[t] changed to 1247.5
ShipConfiguration DistanceFromHullBottom[m] changed to 16.211
ShipConfiguration CentreOfGravity changed to 16.211
ShipConfiguration MastSpace[m] changed to 4.0
ShipConfiguration MiddleDeckSpace[m^2] changed to 187.5
ShipConfiguration Provisions[t] changed to 18.900
ShipConfiguration ElectricalConsumption[kW] changed to 110.5
ShipConfiguration Speed[Kn] changed to 28.628
TotalShipsCompany Weight[t] changed to 4.725
TotalShipsCompany Provisions[t] changed to 18.900
TotalShipsCompany ElectricalConsumption[kW] changed to 94.5
OperationalConstraints ElectricalConsumption[kW] changed to 106.0
OperationalConstraints SafeCentreOfGravity changed to **false**

Thus the ship's centre of gravity has shifted upward from 15.838 metres to 16.211 metres, taking it over the safe level of 16 metres. This causes the SafeCentreOfGravity dimension of component OperationalConstraints to switch to **false**, alerting planners to a possible disaster.

Installing the Deckgun and DeckgunAmmunition components

The Deckgun component was then installed (linked to the Ships-Configuration component by an influence connection), as was the DeckgunAmmunition component. Another ExtraSeaman component was linked to the TotalShipsCompany component. The system now looked like Figure 92.

After a further three Compute time-steps the following changes had been made.

ShipConfiguration Weight[tn] changed to 1322.8
ShipConfiguration DistanceFromHullBottom[m] changed to 16.189
ShipConfiguration CentreOfGravity changed to 16.189
ShipConfiguration UpperDeckSpace[m^2] changed to 15.0
ShipConfiguration MiddleDeckSpace[m^2] changed to 240.0
ShipConfiguration ElectricalConsumption[kW] changed to 130.0
ShipConfiguration TurningMoment changed to 65.0
ShipConfiguration DistanceFromHullBottom[m] changed to 16.189

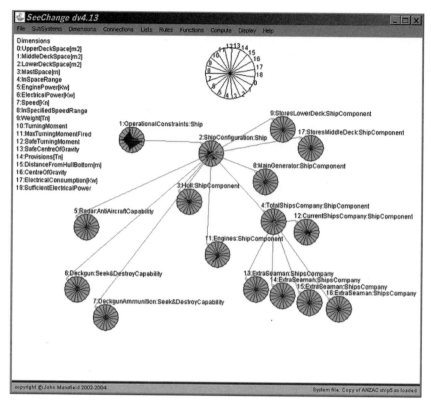

Figure 92.   SeeChange model with the Deckgun and Ammunition installed.

ShipConfiguration Provisions[t] changed to 19.2
ShipConfiguration Speed[Kn] changed to 27.0
TotalShipsCompany Weight[t] changed to 4.800
TotalShipsCompany MiddleDeckSpace[m^2] changed to 130.0
TotalShipsCompany Provisions[t] changed to 19.200
TotalShipsCompany ElectricalConsumption[kW] changed to 96.0
OperationalConstraints ElectricalConsumption[kW] changed to 130.0
OperationalConstraints SafeTurningMoment changed to **false**
OperationalConstraints SufficientElectricalPower changed to **false**
OperationalConstraints SafeCentreOfGravity changed to **false**

Thus, the ship's centre of gravity has shifted downward from 16.211 metres to 16.189 metres, but it is still over the safe level of 16 metres so the SafeCentreOfGravity dimension of the component Operational-Constraints remained at false.

Additionally, the TurningMoment has risen to 65, which is over the safe level of 60, so the SafeTurningMoment dimension of component OperationalConstraints changed to false.

These two conditions make the ship non-operational.

The consumption of electrical power is now above the generated capacity, causing the SufficientElectricalPower dimension of component OperationalConstraints to switch to false, alerting planners to the situation that with this ShipConfiguration it is not possible to operate the Radar component and the Deckgun and Ammunition components concurrently. A difficult situation for an operational warship!

Installing the AuxiliaryGenerator

As the electrical power was now clearly inadequate, the decision was made to install an AuxiliaryGenerator (i.e., introduce an AuxiliaryGenerator component to the system and link it to the ShipsConfiguration component by an influence connection). The system now looked like Figure 93.

After a further three Compute time-steps the following changes had been made.

ShipConfiguration Weight[t] changed to 1324.8
ShipConfiguration ElectricalPower[kW] changed to 140.0
ShipConfiguration Weight[t] changed to 1324.8
ShipConfiguration DistanceFromHullBottom[m] changed to 16.004
ShipConfiguration CentreOfGravity changed to 16.004
ShipConfiguration MiddleDeckSpace[m²] changed to 246.0
ShipConfiguration Speed[Kn] changed to 26.962
OperationalConstraints ElectricalPower[kW] changed to 140.0
OperationalConstraints InSpecifiedSpeedRange changed to **false**
OperationalConstraints SufficientElectricalPower changed to **true**
OperationalConstraints SafeCentreOfGravity changed to **false**

There is now sufficient electrical power so the SufficientElectricalPower dimension changes back to true.

Unfortunately, although the ship's centre of gravity has shifted downward from 16.189 metres to 16.004 metres, it is still over the formal safe

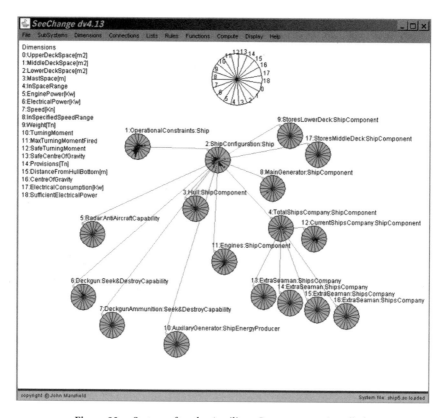

Figure 93. System after the Auxiliary Generator was installed.

level of 16 metres, so the SafeCentreOfGravity dimension of the component OperationalConstraints remains at false. However, it is just over the limit so the ship may be considered operational.

But because the SafeTurningMoment is still false, great caution has to be exercised when deciding to fire the Deckgun.

These various changes, their propagation across the system and their effect on the ship's operational constraints may be seen more clearly in the graphical results shown in Figure 94.

Model of a decision-making component

The simulation of the ship consisted of components that were not inherently in conflict, so the protocol for change relates directly to the environment within the ship. Outside the ship is a wider Naval system

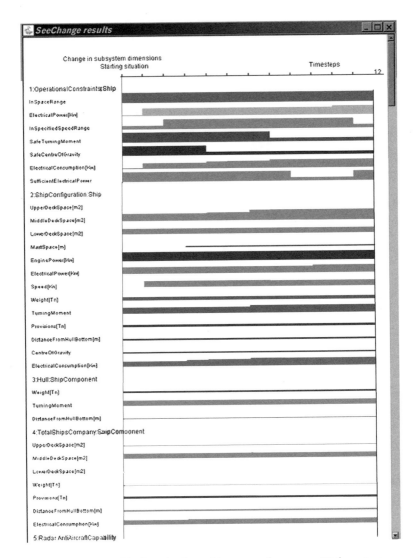

Figure 94. Graphical results from ship system shown over 12 time-steps.

including a number of ships, the Department of Defence, the suppliers, etc. For clarity, this component, wherein the decision to install an auxiliary generator was made, was not included in this simulation.

A model of that component is shown below. The two layers of the wider system are illustrated in Figure 95, as are the influences external to the ship system.

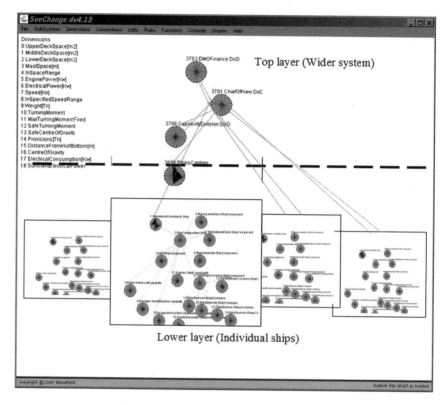

Figure 95. Stakeholders in the wider system.

The stakeholders in this decision are the Ship's Captain, Defence capability organisation, Defence material organisation, Finance, and the Navy Chief. The Navy Chief will take the decision.

Influence

However, in this case, the weights of the influences are not all set at 1.0, the different stakeholders each have a different weight in the decision-making. These different influences are shown in Figure 96. The Chief of Navy is the decision-maker and the decision boils down to two factors:

- Pressure for change: necessary to make the ship operational.
- Resistance to change: extra unbudgeted cost with radar and gun.

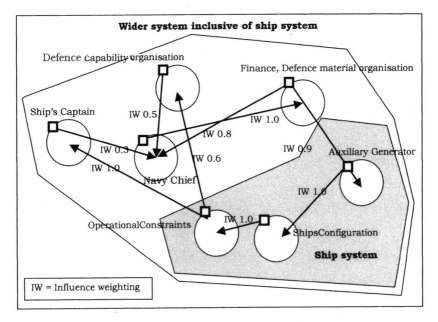

Figure 96. Strengths of influence connections.

The influence weightings of the ship's captain (0.3) and the Defence capability organisation (0.5) on the Navy Chief are relatively small, compared to the influence weighting of the Defence material organisation, Finance (0.8). However the aim of the Navy Chief is to see the radar and gun components installed and operational, and to refuse to install the generator would risk the non-operation of one or other of these systems at critical times. The decision threshold is reached and the decision to install is taken.

Resistance

Turning to another model, that of a resistance to change landscape, we see in Figure 97 the resistance diagram before the decision is made. Even this relatively simple example of the use of SeeChange demonstrated that there was an emergent relationship between the factors affecting the ship's centre of gravity, the factors affecting the ship's turning moment and the ship's seakeeping ability.

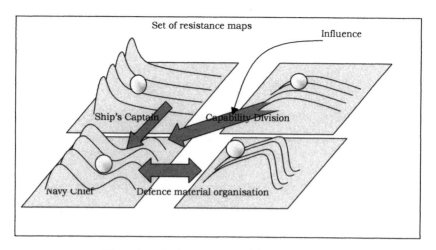

Figure 97.   Resistance maps of the stakeholders.

While this relationship may seem obvious in retrospect, scientists have shown that in large systems like ships, no single person has a complete knowledge of every aspect of the configuration and planned changes for even one ship.

The information about who is going to change what and when exists as static data within a number of different databases, but it is beyond the capability of any single person to constantly correlate the data and conduct a dynamic analysis of the effect of all changes. The results of interference between changes are emergent and occasionally unexpected. Thus a knowledge representation tool is required to make explicit the effect of interfering changes before they occur. For users to trust this tool, it must be open, i.e., not a "black box". What is needed is an industrial-strength tool similar to SeeChange. This tool needs to be an open knowledge representation tool that graphically displays the system using nodes (components) and unidirectional connections. It will show the relationship between nodes, the effect of one node on another, and it will illustrate uncertainty without giving it a quantitative measure. It will also be a system-dynamic tool, in that one of its most important aspects is its temporal facility; this facility allows the user to see the effect of changes over many time-steps.

Another significant point to note is that unlike constraint-based simulations, its nodes will be multi-dimensional and hence the inter-node connection is multi-dimensional. Additionally, many nodes will combine to produce the total environmental effect on a particular node. Any resultant change is also dependent on the intra-node multi-dimensional coupling, requiring a complex trade-off and potentially a multi-dimensional change. The multi-dimensional nature of a node will provide a mechanism to represent the components in the most natural way. The dimensions of the components, the structure of the system and the relationship between nodes (both structural, rule and function-based) are explicit so decision-makers can satisfy themselves on the level of accuracy and assumptions of the model.

If all the information about potential, imminent and ongoing changes is incorporated in such a modelling tool, a simulation may be carried out at an early stage as a "what if" exercise, avoiding the expensive mistakes described in the hypothetical example.

Life is of course not that easy and two major problems would still have to be overcome. First, all the information about planned changes with respect to a single ship and, ideally, to the entire EUREKA class would have to be made available to a single coordinating organisation — in an organisation as large as, say, the Department of Defence this would be difficult to achieve. Second, the volume of data involved for a single ship is immense, so construction of the initial model would probably take two to three years. However, once the rules and functions are defined, subsequent models would take much less time. As I write this, I am not privy to the policies of government departments or defence contractors, so there is the possibility that this has already been done.

12

Real-World Change: Climate

Probably the most discussed topic in the last decade is that of annual changes in the average weather condition around the world. Most people agree that the planet's climate is changing but many disagree on the reason. Massive committees have been created to settle the question but they only create further dissent. Why are we unable to agree on something so vital to our well-being? If you have read this far through the book, it should be clear that the climate is the emergent behaviour of a very complex system that is in constant, internal co-evolution. Such systems should be amenable to simulation, and climate models exist, so why the dissent?

Climate is an exceedingly complex system and I will not attempt to settle the debate on global warming here. What I will do is to show the complexity of the problem and the difficulty of coming up with a definitive answer. First, there are many variables that affect the climate. Second, it is almost impossible to determine the value of all of these variables at a given instant. Finally, establishing the co-evolutionary influences between the known variables is extremely difficult. Let us look first at some of the variables affecting the climate.

Climate Variables

Solar radiation variation

The major source of heat energy on Earth is the Sun, with a minor contribution from geothermal energy. So, variations in the Sun's radiation can have large effects on the Earth's climate. Fortunately, although the Sun is getting hotter as it ages, the change is slow, with the Sun getting about 5% hotter every one billion years. There are shorter-term variations, including the 11-year sunspot cycle, although the mechanism behind these variations is not well understood. Aside from variation of the Sun's energy, other events affect the amount of radiation that arrives at the Earth's surface.

Changes in the Earth's orientation

Despite the classical picture of the Solar System, the Earth does not orbit the Sun in an unchanging circle. In the 1930s, Milutin Milankovitch proposed that the Earth's orbit changed with its cyclic movements, thus varying the amount of energy from the Sun reaching the Earth. His theory claimed that the Earth's orbit was perturbed by three variables — eccentricity, precession and tilt.[134,135]

Eccentricity

The Earth's orbit is an ellipse and its variation from a circle is termed its eccentricity; the more squashed it is, the higher the eccentricity. Because of the effect of the gravitational fields of Saturn and Jupiter, the eccentricity of the Earth's orbit varies between 0.005 and 0.058 over about 423,000 years. Currently, it is about 0.017 (Figure 98). This is important because it

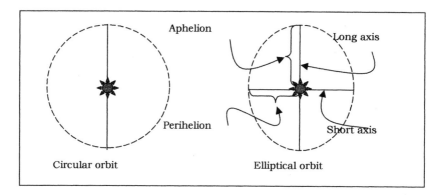

Figure 98.   Eccentricity of the Earth's orbit.

affects the distance of the Earth from the Sun, and therefore the level of solar radiation reaching the Earth and the length of the seasons.

The long axis of the orbit does not vary but the short axis does, changing the eccentricity. The important variable to note is the difference between the aphelion (furthest from the Sun) and the perihelion (closest to the Sun). At the moment, the difference is about 5,000,000 km. This means there is about 7% difference in solar radiation between aphelion (July) and perihelion (January). However, when the orbit is at its greatest eccentricity, the difference is about 23%. This variation affects the length of the seasons. In the northern hemisphere, autumn and winter occur at perihelion when the Earth's movement is fastest, so autumn and winter are shorter than spring and summer. As perihelion occurs during the southern hemisphere's summer and aphelion during the winter, seasons in the southern hemisphere tend to be more extreme than those in the northern hemisphere.

Tilt

As you can see from Figure 99, the Earth's rotational axis is inclined at an angle to a line perpendicular to the plane of the orbit.

This angle varies between 22.1° and 24.5° over about 41,000 years. As the tilt increases, summers receive more radiation and thus are warmer, while winters are colder.

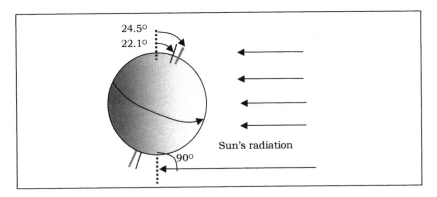

Figure 99. Tilt of the Earth's rotational axis.

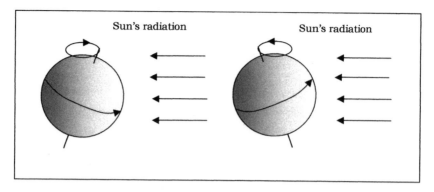

Figure 100.   Precession of the Earth's rotational axis.

Precession

In addition to its tilt, the Earth's axis of rotation moves over about 26,000 years in a circle relative to the stars. As shown in Figure 100, when the axis tilts towards the Sun, the northern hemisphere receives more radiation than the southern hemisphere. When the axis tilts away from the Sun, the reverse happens.

So to start with, we have at least three variables affecting the seasons. These variables combine to form a single complex variation in the radiation reaching the Earth.

Other variables that affect the temperature of the atmosphere more locally are the Earth's albedo, plate tectonics, ocean temperature, air currents, glaciation and polar ice sheets, volcanism, vegetation and human activity.

Albedo

The albedo of the planet refers to the amount of radiation reflected back into space. Cloud, snow, ice or aerosol particles reflect more than deserts or vegetation.[136]

Plate tectonics

On geological time-scales, plate tectonics moves the continents, creates mountain ranges and ocean canyons. It shapes the world and, over long periods, changes the climate. For instance, when the North and South

American plates came together, the Isthmus of Panama was formed, separating the air and ocean currents of the Atlantic and Pacific Oceans.[137]

Ocean variability

Heat is stored in the oceans and is moved around the Earth by means of warm and cold currents. Many climate fluctuations such as the El Niño Southern Oscillation (ENSO) are mainly due to these currents and relatively static volumes of warm and cold water.

Air currents

We are all familiar with air currents, which we call wind. Wind is caused by the movement of air between areas of high and low atmospheric pressure. These "highs" and "lows" drift across the planet under the influence of regional surface temperatures, both land and sea, of solar radiation, which warms the air directly and indirectly, and of the rotation of the Earth. The local atmospheric complex system is the major cause of what we call weather.

Glaciation and polar ice sheets

Glaciers and polar ice sheets are formed when winter snow does not melt in the summer, and fresh snow accumulates on the residue. Over thousands of years, the snow builds up until the pressure turns the lower layers into ice. In the case of glaciers, the ice forms high in the mountains and then slowly flows to lower altitudes due to gravity. Glaciers advance and retreat as the climate becomes colder and warmer; this movement provides a medium-term measure of climate change. Ice sheets, particularly those in Antarctica and Greenland, are more massive than glaciers but are still susceptible to climate change. These areas of ice in turn affect the atmosphere by cooling air above the ice and reflecting solar radiation back into space.

Volcanoes

Volcanoes occur when the tremendous heat and pressure of the magma in the Earth's mantle comes sufficiently close to the surface for it to break through — often catastrophically. When volcanoes erupt, they release into the atmosphere massive amounts of gases and particulate matter ranging from large rocks to very fine dust. The rocks are only of local concern but the dust rises high into the air, forming a pale layer in the upper atmosphere. This layer of dust partially blocks the Sun's radiation, thus cooling the Earth. However, it normally disperses within a year or so. The volcanic gases are another matter; they include sulphides, carbon dioxide and methane, all of which enhance the greenhouse effect and deplete the ozone layer that protects life on Earth from the Sun's harmful UV radiation.

Vegetation

The earliest photosynthetic plants fed on carbon dioxide and excreted oxygen. In doing so, they created an atmosphere conducive to carbon-based life. Modern vegetation continues to lock up carbon in its cells and excrete oxygen. As the plants continue to grow, they maintain a balance that is healthy for other life on Earth. Vegetation affects the composition of the atmosphere and hence the climate, and is itself affected by climate change. Climatic cycles such as El Niño (ENSO) bring rain, enhancing the growth of vegetation, while droughts kill off all forms of vegetation, creating deserts. In good times with heavy rainfall, animals that eat vegetation do well, but in droughts they must migrate or die.

Human influences

When mankind began to cultivate the land and domesticate animals, it started to change the Earth. One example is the clearing of trees to make available more agricultural land, a practice that in some places has led to rising water tables and increasing salinity, eventually reducing the amount of agricultural land.[2]

Since the 18th century, mankind has been using fossil fuels (carbon locked away in ancient plants) in steadily increasing amounts. As these fuels burn, they release carbon dioxide and other gases; gases that have been buried, and out of the atmosphere for millennia. There is a consensus that the emission of these gases, plus waste from industrial processes, deforestation and other human activities, is affecting the composition of the atmosphere.[138] Still, as the worldwide human-produced greenhouse gases result in only about 0.3% of the greenhouse effect,[139] and the CO_2 in the atmosphere has been steadily increasing since the last Ice Age 18,000 years ago,[140] I am doubtful that mankind is the major cause of climate change. However, mankind's water use does influence the local climate. Establishing dams, altering watercourses and wide-area irrigation all modify the terrain and affect the amount of water in the atmosphere. When farmers put in new dams, for example, they can increase the population of wildlife such as kangaroos that then compete with the cattle and sheep for grass — another unforseen consequence.

Atmospheric gases and the greenhouse effect

On reaching the Earth, some 30% of the Sun's high frequency radiation is reflected into space by clouds, dust in the atmosphere and reflective surfaces such as ice sheets. The remaining 70% heats up the land and ocean surfaces, which then re-radiate some of the solar energy into the atmosphere as low frequency heat energy. A layer of gases in the upper atmosphere reflects some of this heat back into the atmosphere. This reflection of heat from the upper atmosphere is called the greenhouse effect because it mirrors the mechanism by which a glass greenhouse is warmed. Jean-Baptiste Fourier noted this atmospheric phenomenon in 1827 and made the greenhouse analogy.[141] The gases that have the greatest effect are water vapour, methane, carbon dioxide and the oxides of nitrogen. This is not all bad, because if the clouds and other greenhouse gases did not exist, the Earth would be covered with ice. As they accumulate, these gases retain more and more of the reflected heat, thus raising the temperature of the Earth.

These are just some of the simple variables that govern the Earth's temperature, and the influence diagram in Figure 101 shows how they might interact. It is clear that the resultant temperature is the emergent behaviour of a complex co-evolutionary system.

Climate Data

Now that we are aware of some of the variables influencing the temperature, how do we determine the value of these variables? In the 21st century

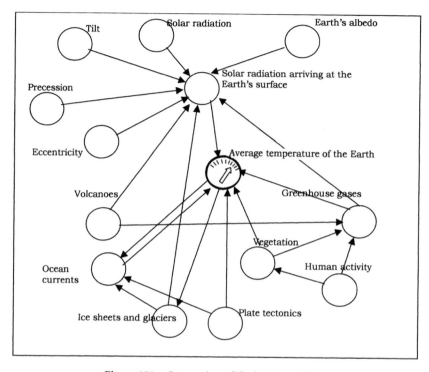

Figure 101. Interaction of the basic variables.

we have thousands of terrestrial sensors taking readings around the planet, and orbiting remote-sensing satellites observing conditions on the Earth's surface and in the atmosphere. Modern communications permit these readings to be accessed by meteorological centres in many countries.

In the recent past, measured in centuries, some records were kept of variables such as temperature, rainfall, wind speed and the like, but these were confined to the land, particularly the cities. Little data was collected from specific spots at sea. Further back, there was no one to keep records, so how do we know what the weather was like thousands and millions of years ago? Without this data, we are not able to model climate trends.

Paleoclimatology

Scientists have devised various ways of looking back into the past via nature's records; collectively, these methods are called paleoclimatology.

They include studying ice cores, tree rings, pollen residues, corals, and sedimentary rocks.

Ice cores

As I noted above, long-term ice sheets are formed when snow accumulates year after year, compressing the lower layers into ice. Each year the snow covers whatever detritus has fallen on the ice during the summer. The snow itself contains tiny bubbles of air that also become trapped in the ice. These annual layers provide a record of weather long ago. To access this record, scientists drill thousands of metres down into ancient ice sheets with long, hollow cylinders like apple corers and bring up samples of ice from way below the current surface. Naturally, they call these samples ice cores. By counting the layers they can determine when each layer was laid down, and by analysing the air in the bubbles they can determine its composition when the snow layer was created. They also analyse the ice to determine the proportion of hydrogen and oxygen isotopes in the water molecules that make up the ice; this proportion varies with the sea temperature at the time the snow fell.

An analysis of the detritus within an ice core layer shows, among other things, the type of dust, volcanic ash or pollen that was falling from the air at the time.

Pollen residues

The study of pollen, particularly fossil pollen, is a science called palynology. Individual plant species have pollen of a distinct shape and texture. When the plant species has been identified, knowledge of the species' growth conditions indicates the climate current at that time.

Tree rings

When a tree trunk is cut through, the growth rings created each year are exposed. Counting the rings determines the age of the tree and some trees have been found to be hundreds of years old. In general, the growth rings are broad in good growing conditions and rainfall, and narrow in poor conditions or drought. So the local climate at a particular time can be estimated by looking at the growth rings.

Corals

Coral shows growth rings in a similar way to trees. In the case of coral, it is the temperature of the water and the salinity that affects its growth.

Rocks

Sedimentary rocks, as their name suggests, were laid down as sediment in lakes and oceans. Dust, earth and gravel from run-off, dead plants and animals large and small drift down to the bottom. Over generations, more and more detritus rains down and compresses the lower layers into rock. Examination of fossil remains in particular strata of sedimentary rock can date the strata, while analysis of the fossils and the rock composition will indicate the climate in which it was laid down.

These and other more arcane methods have enabled scientists to determine the Earth's climate and its trends over hundreds of thousands of years and, in some cases, over millions of years.

Climate Models

As Nigel Gilbert and Klaus Troitzsch said,

> "If we can develop a model which faithfully reproduces the dynamics of some behaviour, we can then simulate the passing of time and thus use the model to 'look into the future'. "[88]

So how do we set about creating a climate model or any model? First, we establish where the system boundary lies, then identify the components within the system and finally determine how each component influences the others.

In this case, we need to set the boundary to include the Earth, the Sun and the larger planets as they all have an effect on our climate. There may be other influences both inside and outside the solar system but for our model we will neglect them, as they are either very small or unquantified.

Some of the components within this system are:

- solar radiation and the components that influence the radiation falling on the Earth, namely the Earth's tilt, precession, orbit eccentricity and albedo;
- ice sheets and glaciers;
- the direction, temperature, salinity, etc., of air and ocean currents;
- volcanic and other geothermal activity;
- plate tectonics, where it disturbs ocean currents and creates mountain ranges;
- greenhouse gases;
- vegetation;
- human activity.

Figure 102 illustrates a descriptive model of these variables.

Assuming that we can measure these variables, the next thing is to determine how one influences the other. This last point has been the subject of

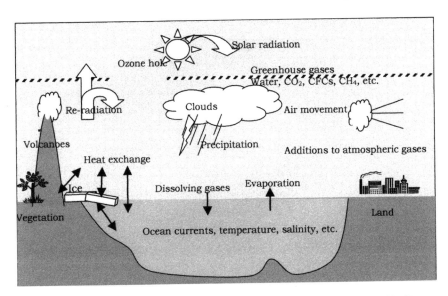

Figure 102.   A descriptive model of some of the variables involved in the Earth's climate.

scientific research for over one hundred years and scientists have devised a number of mathematical equations based on physics and chemistry to quantify the relationships between the components. The simplest, so-called zero-dimensional equation for temperature is:

$$T^4 = \frac{(1-a)S}{4\varepsilon\sigma},$$

where S is the solar constant, i.e., the incoming solar radiation per unit area, currently about 1367 Wm^{-2}, a is the Earth's average albedo, measured to be 0.3, σ is the Stefan–Boltzmann constant — approximately 5.67×10^{-8} J K^{-4} m^{-2}s^{-1}, and ε is the effective emissivity of earth, about 0.612. This gives an average temperature of 15°C (http://en.wikipedia.org/wiki/Climate_model).

I will not inflict them on you but there are also higher dimensional models, radiative-convective models, Earth system models of intermediate complexity and several global climate models each of increasing complexity; the most complex being atmospheric global climate models that combine a sea surface temperature model with an atmosphere model.

Now think back to the influence diagram in Figure 101 above, add to it the complex interactions of the models I have listed and you will get some idea of just how difficult it is to make climate predictions. Additionally, there are many aspects of climate that are not yet included in these models.

Simulation of Climate Models

Now that we have a model, we can incorporate it into a computer program, similar to but vastly more complex than SeeChange, which can then be used to forecast future trends. On the planet, the atmosphere and oceans are divided into a multi-layered grid of cells (Figure 103) and, where possible, data is gathered for each cell. This data includes temperature, humidity, air pressure, and wind speed and direction. The model uses equations from general physics to simulate fluid motion, radiation, convection, evaporation, land and ocean surface heat exchange, cloud cover and albedo.

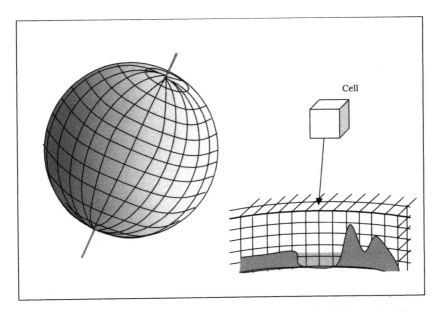

Figure 103. The atmosphere and oceans are divided into a multi-layered grid.

Figure 104 shows how a cell interacts with its six neighbours and how each of those cells interact with their neighbours and so on; creating a co-evolutionary system that ripples round the entire planet. This is a very complex system.

The size of the cell is important too. If it is too large, variations in local temperature, pressure and humidity are averaged out and the simulation becomes too general. As we make the cells smaller, their number balloons; current super-computers can handle between 400,000 and 500,000 cells in a combined atmosphere/ocean model. Each cell, with the exception of those at the extreme top and bottom, has six influences on its neighbours, so if we take the smaller figure of 400,000 cells this works out to be 2.4 million interactions (6 × 400,000). The coupling between the cells is strong for the five attributes of temperature, humidity, air pressure, and wind speed and direction, therefore the simulation is required to calculate some 120 million (5 × 2.4 million) complicated equations for each step. Add to this the calculations for fluid motion, radiation, convection, evaporation, heat exchange, cloud cover and albedo, and we increase this figure substantially.

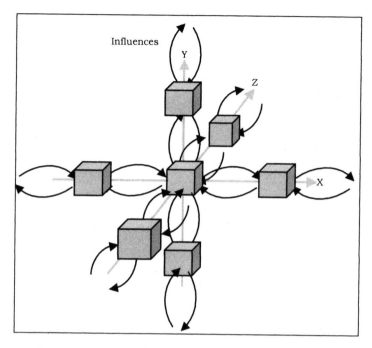

Figure 104.   A cell with its six neighbours in three dimensions.

Remember that to be useful, each step in the simulation must be shorter than real time, preferably much shorter. So we have a trade-off between a quicker, generalised forecast and a slower, more accurate forecast. For example, scientists from Australia's Commonwealth Science and Industrial Research Organisation (CSIRO) report that

"Climate models must approximate processes that are on scales too small to include explicitly, such as the formation of cumulus clouds that may be only a few kilometres wide. ... When models simulate the climate by using only past greenhouse gas changes, the results show surface warming greater than observed. When estimates of the reflective effect of sulphate aerosol (tiny particles suspended in the air, from natural or human-induced sources) are also incorporated, models are better at matching both global average temperature changes and the patterns of the changes to observations. Including solar radiation changes, volcanic aerosol and stratospheric ozone in models further improves the match between the simulation and the observed record."[142]

The difficulty of modelling this complexity clearly reduces the accuracy of the climate simulation and opens it to dispute. As computer capability grows, we can use smaller cells and add more influences to the model.

All this leaves aside the associated co-evolutionary feedback; for example, as the average temperature rises, the Western world turns up its air-conditioners, thus using more fossil fuel and generating more greenhouse gases which may make it warmer still. The co-evolution of very large complex systems always leads to surprises. A rise in atmospheric CO_2 can lead to more acid seas that could, theoretically, lead to the extinction of plankton, which in turn would lead to the extinction of krill, which leads to the death of fish and mammals and hence to starvation on land.

Our climate must rate as one of the most complex systems mankind has ever studied.

13

In the Future

This book has discussed the relationship of a model of co-evolutionary change with the real world of complex systems, I have offered some suggestions in the hope that they may help to alleviate the problem of complex systems failure. The phenomenon of human instigated co-evolutionary change does not only occur in the context of socio-technical systems. It occurs in the intent and effect of laws and regulations. It occurs within organisations and in the marketplace. It occurs in environmental planning and in agriculture. It also occurs in areas without direct human influence such as the climate and micro- and macro-ecologies. Once some thought is given to it, the concept of co-evolutionary change and a protocol for change appears universally pervasive, but relatively little is known about the mechanisms and relationships by which interference between propagating changes operates and affects system failure.

Surprise is Always the Result of Change, Accept It

If you are designing an aircraft, drafting a new law or proposing to make almost any change at all, you need to recognise that the act of changing a complex environment has a high probability of causing unanticipated and unintended behaviour. The result of making a change in a system other than a very simple one is always an emergent behaviour and thus change should not be equated with progress! Change is always co-evolutionary within any environment. Extensive simulation or other forms of impact analysis will reduce but not eliminate the uncertainty. You need to note that environments co-evolve and in evolving they grow.

So what can we do if we accept the uncertain outcomes of any change and wish to exploit them?

- First, use frameworks that facilitate change and future hindsight.
- Next, discard the pretence that the exact requirements of a complex design can be specified in advance and hence, discard the pretence that the design can be fully specified in advance.
- Then, those actually implementing a design need to steer its growth in the direction of the currently desired general goal. Guide it towards the general goal in steps but do not decide on the path in advance because the actual path will go in directions you cannot anticipate.

I hope this book demonstrates that the effect of changes propagating within a complex system is inherently unpredictable and there is a strong probability of the system failing to meet expectations. Unfortunately, there is not and cannot be a single solution for preventing chaotic system failure. However, a simulation program coupled with the alleviative statements above suggests a mechanism for improving the success rate and turning the nature of change to one's advantage.

Bibliography

1. Tegmark, M., Parallel universes, *Scientific American*, **5**, 2003: 37.
2. Lines-Kelly, R., NSW agriculture and biodiversity, *Agnote* DPI-453, July 2003.
3. Leaning Tower of Pisa, *Encyclopædia Britannica*. Retrieved 04/07/2009 from Encyclopædia Britannica Online: http://www.britannica.com/EBchecked/topic/333926/Leaning-Tower-of-Pisa.
4. Kleyn, T. and Jozefowicz, J., Wasteland created by human hands, *Hamburg Evening News*, 1985.
5. Simon, H. A., *The Sciences of the Artificial* (MIT Press, 1969).
6. BMC, *The Bhopal Disaster* (Bhopal Municipal Corporation, 2003).
7. Dominique, L., *Five Past Midnight in Bhopal: The Epic Story of the World's Deadliest Industrial Disaster* (Warner Books, 2002).
8. Damveld, H., IAEA revises report on cause of Chernobyl accident, *WISE News Communique*, 1993.
9. Editorial, The accident at Chernobyl, *Atomic Energy Insights*, 1996.
10. Sewell, L. and Lazo, E., *Nuclear Accidents: Three Mile Island* (Health Physics Society Public and Media Information, 2002).
11. Kemeny, J. G., Three Mile Island: 1979 — Attitudes and Practices, *Report Of The President's Commission On The Accident At Three Mile Island*, 1979.
12. Maher, M. and Poon, J., Modelling design exploration as co-evolution, *Microcomputers in Civil Engineering*, **11**(3), 1994: 195–210.
13. Cross, N. and Dorst, K., Co-evolution of problem and solution spaces in creative design, in Gero, J. and Maher, M. (eds.) *Computational Models of Creative Design IV* (University of Sydney, 1999).

14. Bak, P., *How Nature Works* (Oxford University Press, 1997).

15. Bak, P. and Tang, C., Earthquakes as a self-organized critical phenomenon, *Journal of Geophysical Research*, **B94**, 1989: 15635.

16. Buchanan, M., *Ubiquity. The Science of History ... Or Why the World is Simpler than We Think* (Weidenfeld and Nicolson, 2000).

17. Mardsen, P., Forefathers of memetics: Gabriel Tarde and the laws of imitation, *Journal of Memetics — Evolutionary Models of Information Transmission*, **4**, 2000.

18. Tarde, G., *The Laws of Imitation* (trans. Parsons, E. C.) (Peter Smith, 1962).

19. Ashby, W. R., *An Introduction to Cybernetics* (Chapman and Hall, 1956).

20. Beer, S., *Decision and Control* (John Wiley and Sons Ltd, 1956).

21. Beer, S., *Platform for Change* (John Wiley and Sons Ltd, 1975).

22. Beer, S., *The Heart of Enterprise* (John Wiley and Sons Ltd, 1979).

23. Beer, S., *Brain of the Firm*, 2nd edn. (John Wiley and Sons Ltd, 1981).

24. Beer, S., The Viable System Model: Its provenance, development, methodology and pathology, *Journal of the Operational Research Society*, **35**(1), 1984: 7–25.

25. Beer, S., *Diagnosing the System: For Organisations* (John Wiley and Sons Ltd, 1985).

26. Beer, S., Recursion zero: Metamanagement, *Transactions of the Institute of Measurement and Control*, **14**(1), 1992: 51–56.

27. Beer, S., *Beyond Dispute — The Invention of Team Syntegrity* (John Wiley, 1994).

28. Forrester, J. W., System Dynamics and the Lessons of 35 Years, in Greene, K. B. D. (ed.), *The Systemic Basis of Policy Making in the 1990s* (Massachusetts Institute of Technology, 1991).

29. Forrester, J. W., Industrial dynamics — A major breakthrough for decision makers, *Harvard Business Review*, **36**(4), 1958: 37–66.

30. Brooks, F. P., *The Mythical Man-Month* (Addison-Wesley, 1975).

31. Brooks, F. P., *The Mythical Man-Month: Essays on Software Engineering Anniversary Edition* (Addison-Wesley, 1995).

32. Cropley, D., *Personal communication*, 2003.

33. Mitchell, M. and Newman, M., *Complex System Theory and Evolution* (Santa Fe Institute, 2001).

34. Ropohl, G., Philosophy of socio-technical systems, *Techné: Journal of the Society for Philosophy and Technology*, **4**(3), 1999.

35. Kling, R., When gunfire shatters bone — Reducing sociotechnical systems to social relationships, *Science Technology & Human Values*, **17**(3), 1992: 381–385.

36. Cook, S. C., On the acquisition of systems of systems, *Proceedings of the 2001 INCOSE International Symposium*, Melbourne, 2001.

37. Holland, J. H., *Adaption in Natural and Artificial Systems* (MIT Press, 1992).
38. Holland, J. H., *Emergence: From Chaos to Order* (Oxford University Press, 1998).
39. Casti, J., *Would Be Worlds* (John Wiley, 1997).
40. Axelrod, R. and Cohen, M. D., *Harnessing Complexity: Organizational Implications of a Scientific Frontier* (Free Press, 1999).
41. Bar-Yam, Y., *Dynamics of Complex Systems* (Perseus Books, 1997).
42. Editorial, Lessons of Longford, *Australian Energy News*, 1998. http://www.isr.gov.au/resources/netenergy/aen/aen10/10longford.html.
43. Brown, E., Energy systems expertise is key to critical infrastructure center, Argonne National Laboratory Infrastructure Assurance Center, **17**(2), 1999.
44. Poincare, H., *Les Methodes Nouvelles de la Mecanique Celeste* (Gauthier-Villars, 1892).
45. Rittel, H. and Webber, M., Dilemmas in a general theory of planning, *Policy Studies*, **4**(1), 1973: 155–169.
46. Johnson, J., Chaos: The dollar drain of IT project failures, *Application Development Trends*, 1995: 41–47.
47. Larkowski, K., Latest Standish Group CHAOS Report Shows Project Success Rates Have Improved by 50%, *Report of the Standish Group International, Inc*, 2003.
48. Grayson, I., World congress on IT in Adelaide, *The Australian*, 2002.
49. Janz, C., IT branded "a costly failure", *The Australian*, 2002.
50. Gibbs, W., Software's chronic crisis, *Scientific American*, **271**(3), 1994: 72–81.
51. Mitev, N., Social, organisational and political aspects of information systems failures: The computerised reservation system at French railways, in *Fourth European Conference on Information Systems*, Lisbon, 1996.
52. Benyon-Davies, P., Information systems failure: The case of the London ambulance service's computer aided despatch system, *European Journal of Information Systems*, **4**, 1995: 171–184.
53. Plunkett, S., WestPac abandons high-tech hopes, *Business Review Weekly*, 1991: 26–27.
54. Glass, R. L., Buzzwordism and the epic $150 million software debacle, *Communications of the ACM*, **42**(8), 1999: 17–19.
55. Buckell, J., Back in black but beset by bugs, *The Australian*, 2003.
56. Bajkowski, J., CRM fever blamed for $409 million NAB software write-down, *ComputerWorld*, 2004.
57. Kaihla, P., Inside Cisco's $2 billion blunder, *Business 2.0.*, 2002.
58. Laudon, K. C. and Laudon, J. P., *Management Information Systems* (Prentice-Hall, 1996).

59. Lampson, B. W., On reliable and extendable operating systems. In conference on software engineering techniques, *NATO Science Committee*, Rome, 1969.

60. Voas, J. M., Disposable Information Systems: The Future of Software Maintenance? *Journal of Software Maintenance: Research and Practice*, 11, 1999: 143–150.

61. IEEE, 1219, IEEE Standard for Software Maintenance, *IEEE Software Engineering Standards Committee*, 1998.

62. NASA, *Report of Columbia Accident Investigation Board*, 2003.

63. McIntosh, M. K. and Prescott, J. B., *Report to the Minister for Defence on the Collins Class Submarine and Related Matters*, 1999.

64. Delbridge, A. (ed.), *Macquarie Dictionary* (The Macquarie Library Pte. Ltd., 1991).

65. Albert, R. and Barabasi, A. L., Statistical mechanics of complex networks, *Review of Modern Physics*, 74, 2002: 47–97.

66. Erdős, P. and Rényi, A., On random graphs, *Publicationes Mathematicae*, 6, 1959: 290–297.

67. Newman, M. E. J., Watts, D. J. and Strogatz, S. H., Random graph models of social networks, *Proceedings of the National Academy Sciences of USA*, 99(Suppl. 1), 2002: 2566–2572.

68. Watts, D. J. and Strogatz, S. H., Collective dynamics of "small-world" networks, *Nature*, 393, 1998: 440–442.

69. Watts, D. J., *Small Worlds: The Dynamics of Networks between Order and Randomness* (Princeton University Press, 1999).

70. Anderson, P. W., More is different, *Science*, 177(4047), 1972: 393–396.

71. Devezas, T. C. and Corredine, J. T., The biological determinants of long-wave behaviour in socioeconomic growth and development, *Technological Forecasting and Social Change*, 68, 2001: 1–57.

72. Eisenhardt, K., Making fast decisions in high-velocity environments, *Academy of Management Journal*, 32(3), 1989: 543–576.

73. Eisenhardt, K. and Galunic, D. C., Co-evolving: At last a way to make synergies work, *Harvard Business Review*, 2000: 91–101.

74. Mansfield, J. and Kaplan, S., Designing for co-evolution in information systems, *Proceedings of Conference on Complex and Dynamic Systems Architectures*, Brisbane, Queensland, 2001.

75. Gould, S. J., *Wonderful Life: The Burgess Shale and the Nature of History* (Replica Books, 1998).

76. Thomas, L., *The Medusa and the Snail: More Notes of a Biology Watcher* (Viking Press, 1979).

77. Preface, *Book of Common Prayer*, 1662 edition.
78. Sterman, J. D., System dynamics modeling: Tools for learning in a complex world, *California Management Review*, **43**(4), 2001: 8.
79. Dawkins, R., *The Selfish Gene* (Oxford University Press, 1987).
80. Doyle, J., An introduction to complexity, *The Virtual Engineering and Complexity Workshop*, California Institute of Technology, 1997.
81. Wright, S., Surfaces of selective value, *Proceedings of the National Academy of Science*, USA, 1967.
82. Kauffman, S., *At Home in the Universe: The Search for Laws of Self-Organisation and Complexity* (Oxford University Press, 1995).
83. Kelly, K., *Out of Control: The New Biology of Machines, Social Systems and the Economic World* (Addison-Wesley, 1994).
84. Hannan, M. T. and Freeman, J., Structural inertia and organizational change, *American Sociological Review*, **49**(2), 1984: 149–164.
85. von Neumann, J., The general and logical theory of automata, in Taub, A. H. (ed.) *Collected Works*, Vol. 5. (Pergamon Press, 1963).
86. von Neumann, J., *Theory of Self-Reproducing Automata* (University of Illinois Press, 1966).
87. Still, K., *Crowd Dynamics Limited Web site*, 2003.
88. Gilbert, N. and Troitzsch, K. G., *Simulation for the Social Scientist* (Open University Press, 1999).
89. Levinthal, D. A. and Warglien, M., Landscape design: Designing for local action in complex worlds, *Organization Science*, **10**(3), 1999: 342–357.
90. Sterman, J. D., *A Skeptic's Guide to Computer Models* (Sloan School of Management, MIT, 2002).
91. Gardiner, N., *Complex Systems Summer School — The Long Term Ecological Research Network* (The Santa Fe Institute, Santa Fe, 2003).
92. Simon, H. A., *The Architecture of Complexity, The Sciences of the Artificial* (MIT Press, 1969).
93. Arnott, S. D., Design of a reusable systems of systems architecture for constructive simulation, *SETE2000 Conference*, 2000.
94. Maier, M. W., *Architecting Principles for Systems-of-Systems* (University of Alabama, 2000).
95. Schach, S. R., *Software Engineering with Java* (Irwin, 1997).
96. Zotov, D., Learning the lessons from aircraft accident investigations, *Joint Meeting of SESA-SA Chapter, INCOSE-Australia, and the SEEC Research Group*, Adelaide, 2004.
97. Safe-Roof-Systems-Inc., The Incidence of Roof Collapse, 2004. http://www.senteck.com/need/incidence.html.

98. Page, B. I. and Shapiro, R. Y., *The Rational Public: Fifty Years of Trends in Americans' Policy Preferences* (University of Chicago Press, 1992).

99. Gardner, M., Mathematical games, *Scientific American*, **112**, 1971: 224.

100. Gardner, M., *Wheels, Life and Other Mathematical Amusements* (Freeman, 1983).

101. Berlekamp, E. R., Conway, J. H. and Guy, R. K., *Winning Ways for Your Mathematical Plays*, Vol. 2 (Academic Press, 1982).

102. Wolfram, S., Statistical mechanics of cellular automata, *Caltech preprint* CALT-68-915, 1982.

103. Wolfram, S., Cellular automata as simple self-organizing systems, *Caltech preprint* CALT-68-938, 1982.

104. Wolfram, S., *A New Kind of Science* (Wolfram Media, 2002).

105. Zuse, K., Rechnender raum, *Elektronische Datenverarbeitung*, **8**, 1967: 336–344.

106. Zuse, K., *Calculating Space (Rechnender Raum)* (MIT, 1970), pp. 1–7.

107. Thompson, D. A., *On Growth and Form* (Cambridge University Press, 1961).

108. Langer, J. S., Instabilities and pattern formation in crystal growth, *Review of Modern Physics*, **52**, 1980: 1–28.

109. Mandelbrot, B., *Fractals: Form, Chance and Dimension* (Freeman, 1977).

110. Minsky, M., *Computation: Finite and Infinite Machines* (Prentice-Hall International, 1972).

111. Green, D. G., Cellular automata models of crown-of-thorns outbreaks, in Bradbury, R. H. (ed.), *Acanthaster and the Coral Reef: A Theoretical Perspective* (Springer-Verlag, 1990).

112. Sato, K. and Iwasa, Y., Modeling wave regeneration in subalpine Abies forests: Population dynamics with spatial structure, *Ecology*, **74**, 1993: 1538–1550.

113. Sieburg, H. B. and Clay, O. K., The cellular device machine development system for modeling biology on the computer, *Complex Systems*, **5**, 1991: 575–601.

114. Rocha, L., From artificial life to semiotic agent models, Los Alamos National Laboratory, 1999.

115. Davidsson, P., Multi agent based simulation of socio-technical systems, *MABS 2000: The Second Workshop on Multi Agent based Simulation*, Boston, 2000.

116. Reynolds, C. W., Flocks, herds, and schools: A distributed behavioral model, *Computer Graphics*, **21**(4), 1987: 25–34.

117. McMullin, B., SCL: An Artificial Chemistry in Swarm, Santa Fe Institute Working Paper Number: 97-01-002, Dublin City University, School of Electronic Engineering, Technical Report Number: bmcm9702, 1997.

118. Forrest, S. and Hofmeyr, S., John Holland's Invisible Hand: An Artificial Immune System, presented at the Festschrift held in honor of John Holland, University of Michigan, 1999.

119. Epstein, J. and Steinbruner, M. D., Modeling civil violence: An agent-based computational approach, *Center on Social and Economic Dynamics, Working Paper*, **20**, January 2001.

120. Mansfield, J., The Co-evolution of optimal team behaviour within a team of non-communicating, differentiated agents, *Inaugural Workshop on Artificial Life, AL'01*, Adelaide University, 2001.

121. Morgan, M. G. and Henrion, M., *Uncertainty: A Guide to Dealing with Uncertainty in Quantitative Risk and Policy Analysis* (Cambridge University Press, 1990; reprinted in 1998).

122. Howard, R. A. and Matheson, J., Influence diagrams, *The Principles and Applications of Decision Analysis*, **2**, 1984: 719–762.

123. Chourey, S., Alaska turns to simulation for air training, *Federal Computer Week*, 2004.

124. Lisagor, M., A brainstorming machine. NASA, FAA to use "plug-and-play" approach to test new concepts in traffic management, *Federal Computer Week*, 2002.

125. Miles, D., Civilian leaders witness life-saving simulation training, *American Forces Press Service*, 2004.

126. Ackoff, R., *Redesigning the Future* (John Wiley and Sons, 1974).

127. Pidd, M., *Tools for Thinking* (Wiley, 1969).

128. Royce, W. W., Managing the development of large software systems, *IEEE WESCON*, 1970. Reprinted in *Proceedings of the 9th International Conference on Software Engineering*, 1987: 328–338.

129. US-DOD, *Parametric Cost Estimating Handbook — A Joint Industry/ Government Initiative*, 1995.

130. Lehman, M. M., Human dimensions in successful software development. feedback, evolution and software technology — The human dimension, *ICSE 20 Workshop*, Kyoto, 1998.

131. Brand, S., *How Buildings Learn* (Phoenix Illustrated, 1997).

132. Cropley, D. and Campbell, P., The role of modelling and simulation in military and systems engineering, *Systems Engineering, Test and Evaluation*, Adelaide, 2004.

133. US-Navy, Flotation characteristics, *Naval Architecture Website*, Module 2, Basics, 2004.

134. Milankovitch, M., *Canon of Insolation and the Ice Age Problem* (Alven Global, 1998).

135. Milankovitch, M., *theorie Mathematique des Phenomenes Thermiques produits par la Radiation Solaire* (Gauthier-Villars, 1920).
136. Goode, P. R. *et al.*, Earthshine observations of the Earth's reflectance, *Geophysical Research Letters,* **28**(9), 2001: 1671–1674.
137. NASA, Panama: Isthmus that Changed the World, NASA Earth Observatory, 2008. http://earthobservatory.nasa.gov/Newsroom/NewImages/images.php3 ?img_id=16401.
138. IPCC, Climate Change 2007: The Physical Science Basis (summary for policy makers), A Report of Working Group I to the Fourth Assessment Report of the Intergovernmental Panel on Climate Change, 2007.
139. Global Warming: A closer look at the numbers, 2009. http://www.geocraft.com/WVFossils/greenhouse_data.htm.
140. Global Warming: A chilling perspective, 2009. http://www.geocraft.com/WVFossils/ice_ages.html.
141. Timeline: Climate change, *New Scientist,* 2006.
142. CSIRO, Modelling climate change, *CSIRO Atmospheric Research Greenhouse Information Paper,* 2002.

Index